Hudson Thomson Jay

A Scientific Demonstration of the Future Life

Hudson Thomson Jay

A Scientific Demonstration of the Future Life

ISBN/EAN: 9783337419660

Printed in Europe, USA, Canada, Australia, Japan

Cover: Foto ©berggeist007 / pixelio.de

More available books at **www.hansebooks.com**

A

SCIENTIFIC DEMONSTRATION

OF

THE FUTURE LIFE

BY

THOMSON JAY HUDSON

AUTHOR OF "THE LAW OF PSYCHIC PHENOMENA," ETC.

CHICAGO

A. C. McCLURG AND COMPANY

1895

I Dedicate this Volume

TO

NOEL LAWRENCE ANTHONY,

TO WHOSE KINDLY COUNSEL, ENCOURAGEMENT, AND ASSISTANCE IN ITS PREPARATION I OWE MORE THAN I CAN EXPRESS,

AND WHOSE FRIENDSHIP IS ONE OF THE GREATEST PLEASURES OF MY LIFE.

PREFACE.

NEARLY three years have now elapsed since the publication of my first work, "The Law of Psychic Phenomena," in which I formulated, tentatively, a working hypothesis for the systematic study and correlation of all psychic phenomena. Before venturing to publish that work, however, I had devoted many years to a patient and thorough investigation of the subject, with the view of ascertaining whether any psychic phenomenon had ever been observed and recorded that was inexplicable under the terms of my hypothesis. Not being able to find a record of such a phenomenon, but finding, on the contrary, that every psychic fact furnished a fresh illustration of the correctness of my theory, I ventured upon its publication. Since then I have continued the search, aided by many able reviews and criticisms of my work, the result being that I have been unable to find a fact or an argument that militates against the truth of the hypothesis then formulated.

I have, therefore, felt justified in appearing before the public again, for the purpose of carrying to their legitimate conclusions some of the principles laid down in "The Law of Psychic Phenomena."

That work was devoted almost exclusively to the consideration of the mental characteristics and powers of man as we find him in this life. The present work is devoted to a scientific inquiry concerning his prospects for a future life.

In pursuing this inquiry, I have endeavored to follow the strictest rules of scientific induction, taking nothing for granted that is not axiomatic, and holding that there is nothing worthy of belief that is not sustained by a solid basis of well-authenticated facts. In other words, I have studied the science of the soul precisely as the physical sciences are studied; namely, from an attentive observation, and a systematic classification, of the facts pertaining to the subject-matter. The facts of the soul, as the terminology indicates, consist of what are known as "psychic phenomena." These phenomena have, from time immemorial, excited the wonder and fed the superstitions of all the races of mankind; and it is humiliating to observe that in no age or nation have the superstitions arising from such phenomena assumed a more gross and palpable form than in the last half of the nineteenth century, and in those nations possessing the highest degree of civilization and culture. In the meantime, however, scientists have begun the study of the phenomena with the view of ascertaining something of their nature and proximate cause; and although the study is yet in its infancy, enough has already been learned not only to remove them from the realm of superstition, but to develop the fact that psychic phenomena furnish the only means by which science can solve the problems of the human soul.

The object of this book is to outline a method of scientific inquiry concerning the powers, attributes, and destiny of the soul, and to specifically point out and classify a sufficient number of the well-authenticated facts of psychic science to demonstrate the fact of a future life for mankind.

The earlier chapters are devoted to a review of the principal arguments for immortality heretofore advanced, with the view of showing their invalidity from a scientific standpoint, as well as demonstrating the necessity for a new departure in the methods of treating this the most important problem of human existence. The phenomena of so-called spiritism necessarily come under this category; and for that reason, as well as for the purpose of a correct classification of psychic phenomena, I have felt compelled to devote considerable attention to the refutation of the arguments recently advanced in support of the spiritistic hypothesis. I have also been compelled, in the interest of correct classification, to devote some attention to the psychic phenomena mentioned in the Old Testament.

If my interpretation of these two classes of phenomena runs counter to the opinions of others, spiritists, on the one hand, may derive consolation from the fact that my interpretation of their phenomena leads to the same general conclusion which they have deduced, namely, that man is heir to a future life; and on the other hand, those who hold to the doctrine of plenary inspiration and to the literal interpretation of the Scriptures, will endorse my general conclusions, since they confirm the essential doctrines of the Christian religion, and invest them with a scientific value possessed by no other religion on earth.

In demonstrating the fact of a future life, I have simply analyzed the mental organization of man, and shown that, from the very nature of his physical, intellectual, and psychical structure and organism, any other conclusion than that he is destined to a future life is logically and scientifically untenable.

T. J. H.

WASHINGTON, D. C., Sept. 5, 1895.

CONTENTS.

CHAPTER I.

INTRODUCTORY.

CHAPTER II.

DEFECTIVENESS OF THE OLD ARGUMENTS.

CHAPTER III.

SPIRITISM AND HYPNOTISM.

CHAPTER IV.

SPIRITISTIC PHENOMENA.

CHAPTER V.

SPIRITISTIC PHENOMENA (*continued*).

CHAPTER VI.

ANCIENT PSYCHIC PHENOMENA.

CHAPTER VII.

ANCIENT PSYCHIC PHENOMENA (*continued*).

CHAPTER VIII.

THE ADVENT OF JESUS.

CHAPTER IX.

THE INTUITIVE PERCEPTION OF TRUTH.

CHAPTER X.

PSYCHIC PHENOMENA OF PRIMITIVE CHRISTIANS.

CHAPTER XI.

MODERN PSYCHIC PHENOMENA.

CHAPTER XII.

HAS MAN A SOUL?

CHAPTER XIII.

HAS MAN A SOUL? (*continued*).

CHAPTER XIV.

HAS MAN A SOUL? (*continued*).

CHAPTER XV.

DUALITY DEMONSTRATED BY ANATOMY.

CHAPTER XVI.

DUALITY DEMONSTRATED BY EVOLUTION.

CHAPTER XVII.

THE DISTINCTIVE FACULTIES OF THE SOUL.

CHAPTER XVIII.

FACULTIES BELONGING TO A FUTURE LIFE.

CHAPTER XIX.

THE DYNAMIC FORCES OF THE MIND.

CHAPTER XX.

THE AFFECTIONAL EMOTIONS OF THE SOUL.

CHAPTER XXI.

PRACTICAL CONCLUSIONS.

CHAPTER XXII.

LOGICAL AND SCIENTIFIC CONCLUSIONS.

A SCIENTIFIC DEMONSTRATION

OF THE

FUTURE LIFE.

CHAPTER I.

INTRODUCTORY.

Bacon's Monument to Common Sense. — The First to recognize the True Value of a Fact. — The Law of Correct Reasoning. — Its Simplicity. — The Essentials of a Correct Hypothesis. — Inductive Reasoning. — The Copernican System. — Defective Methods of Reasoning employed by the Greek Philosophers. — Speculative Philosophy subject to the Law of Reaction. — The Inductive Sciences insure Permanent Progress. — Natural Theology at a Standstill. — The Conflict between Religion and Science. — Voltaire and Paine. — Their Assaults upon Dogma. — Their Religion. — The Triumph of Science. — The Doctrine of Evolution. — A New Controversy. — Religion and Science not Antagonistic. — Immortality a Proper Question for Scientific Investigation. — If True, it is Important. — If Important, it can be Demonstrated.

"MAN, the minister and interpreter of Nature, does and understands so much as he may have discerned concerning the order of Nature by observing or by meditating on *facts: he knows no more, he can do no more.*"[1]

These words are Bacon's; the italics are mine.

If the great Lord Chancellor had written and expounded but that one sentence, he would have been entitled not

[1] Novum Organum, book i. p. 1.

only to the eternal gratitude of all mankind, but to the credit of having builded the grandest monument to Common Sense that was ever erected by human genius. This eulogium will not seem extravagant when it is remembered that Bacon was the first man who taught the world the true value of a fact; that is to say, he was the first to discover and formulate the fundamental truth that all successful inquiry concerning the order of Nature must of necessity be founded upon a solid basis of well-authenticated facts. When we contemplate the wondrous civilization of ancient Greece and Rome, their advancement in the science of government, the beauty and grace of their literature, the subtleties and refinements of their philosophy, the transcendent genius of their artists, the grandeur and nobility of their architecture, it seems strange, incomprehensible, incredible, that the discovery of this self-evident truth was left for a civilization built upon a soil which was not rescued from barbarism when the Parthenon began to decay and the Coliseum to crumble. But such was the tardiness of human progress — the conservatism of the human mind — in the days before it had broken the shackles of authority, when opinions had the force of enactments, and dogmas were regulated by statute. What is now, to the unperverted mind of the average school-boy, a self-evident proposition, struck the scientific mind of the Elizabethan age with the force of a revelation; and it is safe to say that the world owes all its subsequent progress in material science to the process of reasoning and of scientific investigation formulated and developed by Francis Bacon. Nay, more. The world not only owes all its substantial progress to that source, but the inductive process is the sure guaranty of the stability of our civilization, and of its constant advancement for all time.

The laws of correct reasoning are as immutable as the law of gravity; and, properly applied, are as certain and exact in their results as a law of mathematics. They are the natural laws of the human intellect; they are inherent in its nature and constitution. But what is true of every law of Nature is also true of the law of reason; namely, that until it is discovered and formulated by man, he is not in a position to avail himself of its uses, or to reap the benefits of its beneficence. Like every other law of Nature, when once comprehended the law of correct reasoning was found to be simple to the last degree. It is well stated in the opening sentence of the "Novum Organum," and quoted at the beginning of this chapter. It may be restated thus: *Nothing can be known with certainty except by an appeal to facts.* This is inductive reasoning.

Broadly speaking, there are but two methods of reasoning; namely, induction and deduction. The former consists in reasoning from particulars up to generals, and the latter in reasoning from generals down to particulars. Each is proper in its legitimate sphere; but all conclusions depend for their validity upon the correct employment of each in its proper domain, by which one is never allowed to take the place or usurp the functions of the other.

Inductive reasoning, then, consists in observing, verifying, and classifying all the facts attainable pertaining to the subject-matter undergoing investigation, with a view of arriving at the general principle or law which underlies all the observable phenomena. This is the first great step in the process, without which man can never be certain that he knows anything. The utmost care, therefore, is necessary in this step in order to avoid the pitfalls which beset the pathway of every honest investigator. The first of these pitfalls is inaccurate observation; the second is insufficient verification; and the third is the constant tendency of the human mind to generalize from an insufficient number of facts. There are many other sources of error which beset one who would conduct a scientific investigation; but as it would be foreign to the purpose of this book to discuss the subject in detail, I will content myself by pointing out one that does not seem to have attracted its due meed of attention.

Referring to the general tendency of the mind to generalize from an insufficient number of facts, — a propensity which also includes inaccurate observation and insufficient verification, — it will be observed that there is also a tendency to range facts into factions, and to determine general principles by suffrage. This often happens after an investigator has committed himself to an hypothesis. He soon finds that his theory is contradicted by some of his facts, but he consoles himself with the reflection that the majority of his facts sustain his hypothesis, and he triumphantly quotes the old maxim that "Exceptions prove the rule." No more pernicious and fatal error can be entertained. There are no exceptions to the operations of a law of Nature. There exceptions do *not* prove the rule. This maxim holds good only in its application to human laws. It is applicable to them because it often happens that a rule of common law which applies with substantial justice to a great majority of cases, will work irreparable wrong in an exceptional case. Hence courts of equity are established "for the correction of that wherein the law, by reason of its universality, is deficient."[1] But Nature's laws require no courts of equity to provide for exceptional cases. Exceptions prove the rule in human enactments in that they provoke attention to the rule and thus give it emphasis by antithesis. In case of an apparent exception to a supposed law of Nature, one of the two propositions must be true: 1. If it is truly a law, the exception is only apparent, and fuller investigation will demonstrate that fact, and thus emphasize the rule; 2. On the other hand, if one fact refuses

[1] Blackstone.

to range itself under the terms of a supposed law, that fact demonstrates the invalidity of any hypothesis.

Particular stress is laid upon this point for the reason that, as before remarked, it seems to have been lost sight of in many quarters where one would expect to find the strictest rules of scientific investigation rigidly enforced. Newton fully appreciated the weight and importance of the distinction, as is shown by the fact that he long delayed the publication of the "Principia," because of the apparent refusal of one phenomenon to submit to the terms of his hypothesis; and not until it was demonstrated by subsequent discovery that the apparent exception did not exist, did he venture to give to the world the theorem which made his name immortal.

Having established a general principle or law by induction, the process of deduction begins; and if no fact remains to negative the principle, we can take our stand upon the constancy of Nature and the immutability of her laws, and confidently explain the past and predict the future. And this is the test of the correctness of an hypothesis, — that it enables one skilled in the science to which it appertains to predict correctly, to state with scientific certainty what will happen under a given state of circumstances. Thus a knowledge of the laws pertaining to the movement of the heavenly bodies enables the astronomer to predict the phases of the moon and the eclipses with mathematical exactitude. We may take the science of astronomy as an illustration of the processes of inductive reasoning and of all scientific investigation. By the accurate observation of facts for a long series of years by many and independent observers; by comparison of the results of their observations, and by a system of checking, tabulating, verification, and revision constantly employed, aided by the genius of such men as Kepler and Newton, the Copernican system of astronomy was finally wrought out, and the laws governing planetary motion were formulated. This was induction, — reasoning from particular facts up to general principles or axioms. By deduction, the astronomer, taking as his premises these general principles thus established (the constancy of Nature being always assumed), is enabled to explain all the salient features of planetary motion, and to predict with unerring accuracy the phenomena of the future. On the other hand, the Ptolemaic system, which preceded the Copernican, may serve as an illustration of the defective methods of the ancients, arising from inaccurate observation, insufficient verification, and premature generalization.

It must not be understood that, because Bacon was the first to discover and formulate the law of inductive reasoning, he was the first to reason inductively. Men had always reasoned by that method, more or less. Nor must it be inferred that, because he was the first to discover and make known the true value of a fact as an element of logic, he was the first to employ facts as a basis of reasoning. The first man who ever observed the sun rising at one point of the compass and setting at the opposite, observed three hundred and sixty-five facts every year, from which he reasoned inductively up to the general principle, that the sun rises in the east and sets in the west; and he was enabled to predict, from day to day, that the sun would continue so to rise and set. It so happened that the man was approximately right, having observed a sufficient number of facts to justify his belief. But the same man, doubtless, was equally certain that the earth was flat, and that his horizon marked the boundaries of the habitable world. In this he was wrong, and his error arose from defective observation of an insufficient number of facts. Nor in this was he alone. His defective methods of reasoning, differing only in degree and not in kind, were shared by all his contemporaries, and by all his successors, great and small, down to the days of Plato and Aristotle, and from Plato and Aristotle down to the days of Queen Elizabeth.

Until that time all men reasoned by defective methods; for the fundamental law of reasoning had not been discovered. Hence the wisdom of a Socrates or a Plato afforded no protection against the fatal error of deducting the most momentous conclusions from assumed premises; nor could the logic of Aristotle, which, as Bacon declares, "corrupted natural philosophy," prevent him from "constructing the universe out of his Categories."[1] The wisdom of the Greeks, according to Bacon, was disputatious; their science was spectacular; their history was composed largely of tales and rumors of antiquity, and they were always more intent on founding sects and systems of philosophy, and fighting for supremacy in wrangling, than zealous in their search for truth. Their teachings, therefore, often seemed to justify the charge of Dionysius against those of Plato, — that they were "the words of idle old men to inexperienced youth;" and of the Egyptian priest who said of the Greeks that "they were ever children, and had neither antiquity of knowledge nor knowledge of antiquity;" and of Bacon, who, quoting the above, added, "And surely in this they are like children, — they are ready to chatter, but cannot beget."

Nevertheless, no one can fail to appreciate the subtlety of their philosophy, the vigor of their intellects, or the virility of their manhood, whatever may be said of the soundness of their methods of searching for truth. In spite of defective processes of reasoning they have bequeathed to posterity an immortal literature, a deathless fame, and a philosophy which, in many instances, demonstrates an intuitive perception of truths which modern science can only illustrate and confirm. But of true science they had nothing worthy of the name. Like their philosophy, it was

[1] Novum Organum.

speculative, and hence was unable to withstand those ever-present reactionary forces which impel the human mind to rebel against any system of science, philosophy, or belief not based upon observable phenomena or demonstrable propositions.

Hence it was that all the learning and the philosophy, all the arts and the civilization of ancient Greece and Rome could not avert their decadence, nor rescue the intellectual world from the dismal horrors of the long night of mediæval barbarism. It is a common remark that the physical effeminacy of the people of ancient Rome, resulting from the luxurious habits engendered by the refinements of their civilization, rendered them an easy prey to the hordes of vigorous barbarians of Northern Europe, and was thus the primary cause of their downfall. Other instances exist where ancient civilizations have risen and flourished and fallen. Every year fresh discoveries are made of the remains of prehistoric civilizations which must have been in decay, if not extinct, long before tradition began. And in every case, historic or prehistoric, there exist evidences that their extinction was the result of practically the same causes as those which led to the downfall of the Roman Empire. From these facts it has been argued that there must exist a natural law pertaining to civilization analogous to the law of organic nature; namely, that growth results in maturity, maturity in degeneracy, and degeneracy in disintegration, — in other words, that the law of human development is not the law of constant progress, but that civilization moves in successive cycles. Such reasoners look with gloomy foreboding upon the present state of progress in science and the arts as a sure precursor of the imminent decadence of those nations who have attained the higher civilization, and of their ultimate relapse into barbarism.

I cannot so interpret the history of mankind. Our present civilization is built upon a radically different foundation from that of any of the nations whose history may be cited as a precedent. The difference may be illustrated by a single reference. Taking Greece as an example presenting the most striking contrast between the highest degree of her enlightenment and the lowest degree of her degeneracy, the most obvious fact pertaining to the character of her civilization is this: that in not one of the arts or sciences in which she excelled the most barbarous nations which surrounded her was there a single element of power that could give promise of national perpetuity, or even of substantial national progress. The Greeks excelled in philosophy, but it was almost purely speculative, and was therefore subject to the law of reaction. Their science was as speculative as their philosophy, and subject to the same law. They excelled in mathematics, but in the absence of other sciences, of which mathematics is but the handmaiden, it was not an element of power. They excelled in art and in literature, but in neither was there an element of national strength; for though the art of Phidias has never been surpassed, and Homer's rank after the lapse of ages is unchallenged, the sculptor's chisel and the poet's tablet were poor weapons of defence against the superior physical force of their enemies.

On the other hand, the civilization of the present day is founded upon the inductive sciences. In the inductive sciences the law is that of eternal progress. In them there is no possible element of reaction. A proposition or principle of natural philosophy, once established, is as firmly fixed as a proposition in mathematics, and is never afterwards disputed. Every step, therefore, is a step in advance. Every new demonstration of a law of Nature furnishes the basis for a fresh start in a thousand different directions. There is, therefore, no possibility that, either in the purely demonstrative or in the purely experimental sciences, the world can ever again go backward, and there is as little probability that it will ever stand still.

The inductive sciences have within themselves the inherent principles of perpetuity and of progress. Not only that, but they are constantly providing external defences against assaults by physical force. No hordes of barbarians can now swoop down upon a superior civilization, and conquer its people by means of mere muscular superiority; for the inductive sciences have provided appliances which confer upon intelligence and skill a vast superiority over combined muscular and numerical strength, though the latter may be inspired by the most desperate physical courage. Science, and not muscle, is now the prime factor in the struggle of nations for supremacy; for *the victories of war, as of peace, are organized in the laboratories of the inductive sciences.*

The obvious inference is that, other things being equal, so long as the world is under the dominion of the inductive sciences, no civilized people can ever again be conquered except by the agents of a higher civilization.

It is unnecessary to dwell further upon the obvious importance of the discovery which Bacon made; and my only excuse for reciting the *a b c* of the processes of induction is that it is always proper, and frequently important, in the discussion of any question, to recur to fundamental principles. Besides, whilst there is no law of nature more simple, or more easily comprehended, than the fundamental law of human reason, yet there is none that is more habitually and persistently disregarded and set at defiance. It is safe to say that nine-tenths of all that mankind believes, or thinks it believes, is destitute of any solid basis of fact. It is, perhaps, not so much the fault as the misfortune of humanity that this is true. We must not forget that, much as mankind has achieved in the way of wresting from Nature the secret of her laws, the intellectual world is yet in its infancy. It is less than three centuries since man began to comprehend the first principles pertaining to the power which enables him to make an intelligent search for truth. Gigantic strides have been made within that time, it is true; but they have been in one direction only. The material universe has been explored, the dynamic forces of Nature have been enslaved, and the physical condition of man has been ameliorated.

But many problems still remain unsolved which are of far greater importance to mankind than any that have yet yielded to the processes of induction; and they are problems upon which none of the physical sciences throw the faintest glimmer of light.

Natural theology stands precisely where it did when Thales philosophized and Simonides sang; and the arguments are identical with those which Socrates employed in his confutation of the atheism of Aristodemus. Not one of the physical sciences in which we excel the Idumeans has advanced us one step in the solution of the great problem propounded by Job, — "If a man die, shall he live again?"

Indeed, the discoveries of modern science seem to have weakened, rather than strengthened, the old arguments employed to prove the existence of Deity or the doctrine of immortality. Modern physical science has at least weakened the hold which those beliefs had upon humanity; for the scientific mind is prone to hold that what is not proved by induction is, to a certain extent, disproved. And no scientist has ever attempted to demonstrate either of those propositions by induction. Bacon himself does not seem to have regarded theology, natural or revealed, as being susceptible of being brought within the domain of science. On the contrary, he appears to have regarded the essential doctrines of religion as sufficiently well established by revelation. He warns his readers, however, against "an unwholesome mixture of things human and divine," and advises them to "render to faith the things that are faith's."

It does not seem probable that Bacon, whose mind was cast in a severely logical mould, could have overlooked the wide discrepancy between the methods of reasoning which he taught, and those which were at that time necessarily employed in sustaining the fundamental doctrines of religion. Nor does it seem possible that he was insensible to the difficulties which must environ the Church when it should be called upon to defend its faith against the assaults of scepticism, armed with the weapons which he created. Be that as it may, it was not until many years after Bacon wrote that the secondary effects of his philosophy became manifest. As soon, however, as the students of material science became imbued with his wisdom, and began to apply the severe rules of his logic to the investigation of the problems of the physical universe, they began to inquire why the same rules were not applicable to things spiritual; and as soon as it was prudent to do so, they began to demand that the theologian should give as good reasons for the faith that was in him as were required of the scientist for the elucidation of the simplest propositions in natural philosophy. It is needless to remark that this demand has not yet been met with an adequate reply, although the Church has been engaged, with a zeal entirely disproportioned to its success, in defending its strongholds.

It was not, however, until after the beginning of the present century that the real battle between science and religion took a definite form, or that science assumed a seriously threatening aspect towards the fundamental doctrines of religion. It was not until within the memory of men now living that scientists, worthy of the name, became the aggressive opponents of the doctrine of a future life, or attempted to disprove the existence of Deity. The great conflict between religion and science, previous to that time, which may be said to have been begun in the destruction of the Alexandrian Library and ended with the Inquisition, was waged on entirely different grounds. Thus, when Hypatia was stripped naked in the streets of Alexandria by Cyril's mob of monks, dragged into a church, and there killed by the club of Peter the Reader, it was for the offence of teaching mathematics and the philosophy of Plato and Aristotle. The subsequent conflicts were principally respecting such questions as the nature of the Godhead, the nature of the soul, the nature of the world, the age of the earth, the criterion of truth, and the government of the universe. For many hundreds of years these questions were discussed, the principal arguments employed against science being feebly typified by those of Cyril against Hypatia. Even as late as the eighteenth century the religious polemics of the day were not directed against the fundamental truths of natural religion, but against the system of theology which is based upon the interpretation which the priesthood has given to revelation. The works of Voltaire and of Paine may be cited as the best known examples. Each of these writers has been stigmatized as an atheist: but Voltaire believed in God, and steadily upheld the truths of natural religion; whilst Paine, were he living to-day, would find congenial employment in the Unitarian pulpit. The effect of their polemics was great in their day and generation, but it was not lasting. They shook the foundations of creed and dogma, but not of religion. They were not atheists themselves, yet it cannot be denied that their writings have been instrumental in converting many to atheism who have not been able to distinguish between dogma and religion. This effect, however, in the very nature of things, could not be permanent; for no argument not based upon scientific induction can long prevail against the instinct of worship which is inherent in the human mind, or that hope of a life beyond the grave which springs eternal in the human breast.

The science of the nineteenth century, however, has developed an entirely new aspect of the question. The conflict between religion and science still goes on; but the questions are different and the weapons are not the same. It is no longer a question of geography, or of astronomy, or of the shape of the earth, or of its relative magnitude and importance as compared with the other planets in the solar system. All these questions have been settled, and it will not be denied that in each of these conflicts the palm of victory has been awarded to science.

The doctrine of evolution has now given rise to another controversy (it can no longer be called a conflict) between science and religion, or, rather, between scientists and a portion of the Christian Church. On its face it is a controversy relating to the creation and government of the world, — whether it was by a special creative act of God, followed by incessant divine intercession, or by the operation of primordial and immutable law. The Church, however, is by no means united in its opposition to the doctrine of evolution. On the contrary, many of its most progressive and enlightened adherents accept the doctrine without qualification, whilst others attempt to harmonize it with the Mosaic account of creation. There can be little doubt of the ultimate triumph of science in this, as in other controversies; and there can be as little doubt that, when the day of its triumph comes, it will be found that true religion has lost nothing. Religion has never lost anything as a result of the triumphs of science, but only as a result of misdirected zeal in opposing science. Religion, therefore, has nothing to fear from the doctrine of evolution, or from any other science, if religion is truth; for no truth is inconsistent with any other truth.

The real danger consists, not in the conflict of religion with science, but in the failure of the Church to meet the demands of science. The latter reaches its conclusions from the observation of facts, and holds that nothing is worthy of belief that is not sustained by observable phenomena; and it demands of the Church the same quality and character of evidence of what that institution claims to be truth as is demanded of science in support of its propositions. The failure to meet this demand is filling the civilized world with materialism; for scientists are prone to hold that whatever is not susceptible of scientific proof by the processes of induction is, *ipso facto*, disproved. On the other hand, this proposition is offset by many of the clergy by the declaration that questions relating to immortality and the existence of a God are not proper subjects of scientific investigation; that spiritual truths must be discerned by spiritual perception, — must be seen by the eye of faith alone, — and are necessarily undemonstrable by scientific induction. Herein lies the fundamental error, — an error which is fast driving the scientific world into the ranks of materialism; for science holds that truth is only sacred in the sense that error should never be allowed to usurp its place, and that anything which man desires to know is a legitimate subject of scientific investigation. In this declaration science is undoubtedly right; and it might well go a step farther, and declare that *anything which it is important for man to know can sooner or later be scientifically demonstrated by the processes of inductive reasoning.* In making this declaration I make no distinction between physical and spiritual laws. A psychic fact is just as much a fact as a granite mountain. If there is a God, it is important for man to know it; and there are facts which will prove it. If there is a life beyond the grave, it is important for man to know it; and there are facts which will demonstrate it beyond a peradventure. It is to the task of presenting a few of these facts that I address myself in succeeding chapters.

CHAPTER II.

DEFECTIVENESS OF THE OLD ARGUMENTS.

The Four Leading Arguments: 1. Analogical Reasoning inherently Defective. — Metamorphosis. — Butler's Analogy. — Physical Laws not Identical with Spiritual Laws. — Illustration is not Proof. — Averroism. — Emanation and Absorption. 2. Prescriptive Authority. — The Hiding-Place of Power. — The Priesthood and Divine Revelation. — Inductive Arguments of the New Testament. 3. Philosophical Speculation. — Emerson's Belief. — His Despair of Proof. — Plato's Phædo. — His Three Arguments for Immortality. — The Doctrine of Contraries. — Reminiscence. — Reincarnation. — The Capacity of Great Men for Minute Subdivision. — The Soul a Simple Substance. — The Phædo a Promoter of Suicide. 4. Instinctive Desire. — A Valid but not Conclusive Argument.

BEFORE proceeding with the line of argument which it is proposed to adopt in the discussion of the subjects under consideration, I deem it proper to say a few words regarding the methods of reasoning which have heretofore prevailed, with the view of pointing out a few of the salient defects in the arguments commonly employed, as viewed from a purely scientific and logical standpoint. This will not be done in any spirit of censure or fault-finding; for I cannot be unaware of the difficulties which have heretofore environed the whole subject-matter, and of the practical impossibility of formulating a conclusive argument in the absence of those facts which have come to light only within the last quarter of a century. No one can justly be blamed for failure to reason inductively in the absence of facts pertaining to the subject-matter of his speculation; and no man can be justly censured, except from an ultra-scientific standard of reasoning, for accepting, without too critical an examination, such arguments as were available in support of a doctrine which has given to mankind so much of comfort and consolation as the belief in a future life has afforded to a great majority of the human race. For, much as we may deprecate many of the dogmas of the Church, much as we may deride the crude speculations of men regarding the future destiny of the soul and its rewards and punishments, the fact remains that they have all served their purpose in their day and generation; and it is difficult now to see how the world could have gotten along without them. Their terrors have been a potent means of restraint from wrong-doing among men whom nothing else could restrain; and their promises have filled the human heart with consolation in this life, and placed the iris above the door of the sepulchre. Each dogma, each system of religious belief, has been a step in the evolution of the human mind towards a knowledge of the attributes, the powers, and the destiny of man.

In looking backward, therefore, over the tortuous and difficult pathway which the human mind has been compelled to tread in its search for evidences of the reality of that most important of all the objects of human aspiration, immortal life, it would ill become us to despise, or affect to despise, any one of the gradients by which mankind has been gradually lifted into a purer intellectual atmosphere, and enabled to enjoy a clearer perception of truth. In this spirit it is proposed briefly to examine the arguments which have heretofore been advanced in support of the doctrine of a future life, and to test their validity by the simple but infallible rules of logic which every intelligent reader understands and appreciates. If the old arguments are found invalid or inconclusive from a scientific standpoint, it will then be in order to inquire what science has to offer in their place.

In order that I may not be accused of misstating the fundamental grounds upon which mankind has built its hopes of a life beyond the grave, I quote the following passage from Alger's admirable work, in which is summarized the "suggesting grounds on which the popular belief rests":

> "When, after sufficient investigation, we ask ourselves from what causes the almost universal expectation of another life springs, and by what influences it is nourished, we shall not find adequate answer in less than four words: feeling, imagination, faith, and reflection. The doctrine of a future life for man has been created by the combined force of instinctive desire, analogical observation, prescriptive authority, and philosophical speculation. These are the four pillars on which the soul builds the temple of its hopes; or the four glasses through which it looks to see its eternal heritage."[1]

These being the "four pillars" on which the temple is built, it is obvious that if either one of them is found to rest upon an insecure foundation, the whole structure must be in danger; and if all are found to have been built upon logical quicksands, the superstructure must inevitably fall. Dropping the architectural simile, it must be said of the four grounds of belief that some of them embrace valid arguments, but none of them are conclusive. The first in the order named — "instinctive desire" — also stands at the head in point of validity. Its discussion, however, will be reserved for the last of the series, for reasons which will be obvious when it is reached.

The question of "analogical observation" will first receive our attention, although a large part of that which comes naturally under the head of "philosophical speculation" must also be included under this head. I cannot sum up the leading analogical arguments in favor of an immortal

[1] Critical History of the Doctrine of a Future Life, p. 38.

life in better language than by quoting again from the same author:[1] —

"Man, holding his conscious being precious beyond all things, and shrinking with pervasive anxieties from the moment of destined dissolution, looks around through the realms of nature, with thoughtful eye, in search of parallel phenomena further developed, significant sequels in other creatures' fates, whose evolution and fulfilment may haply throw light on his own. With eager vision and heart-prompted imagination he scrutinizes whatever appears related to his object. Seeing the snake cast its old slough and glide forth renewed, he conceives so in death man but sheds his fleshy exuviæ, while the spirit emerges, regenerate. He beholds the beetle break from its filthy sepulchre, and commence its summer work; and straightway he hangs a golden scarabæus in his temples as an emblem of a future life. After vegetation's wintry deaths, hailing the returning spring that brings resurrection and life to the graves of the sod, he dreams of some far-off spring of humanity, yet to come, when the frosts of man's untoward doom shall relent, and all the costly seeds sown through ages in the great earth-tomb shall shoot up in celestial shapes. On the moaning seashore, weeping some dear friend, he perceives, now ascending in the dawn, the planet which he lately saw declining in the dusk; and he is cheered by the thought that

'As sinks the day-star in the ocean-bed,
And yet anon repairs his drooping head,
And tricks his beams, and with new-spangled ore
Flames in the forehead of the morning sky,
So Lycidas, sunk low, shall mount on high.'

"Some traveller or poet tells him fabulous tales of a bird which, grown aged, fills his nest with spices, and, spontaneously burning, soars from the aromatic fire, rejuvenescent for a thousand years; and he cannot but take the phœnix for a miraculous type of his own soul springing, free and eternal, from the ashes of his corpse. Having watched the silkworm, as it wove its cocoon and lay down in its oblong grave apparently dead, until at length it struggles forth, glittering with rainbow colors, a winged moth, endowed with new faculties and living a new life

[1] Op. cit., pp. 38, 39.

in a new sphere, he conceives that so the human soul may, in the fulness of time, disentangle itself from the imprisoning meshes of this world of larvæ, a thing of spirit beauty, to sail through heavenly airs; and henceforth he engraves a butterfly on the tombstone in vivid prophecy of immortality. Thus a moralizing observation of natural similitudes teaches man to hope for an existence beyond death."

From time immemorial the metamorphosis of the caterpillar into the butterfly has been used as a standard illustration of the thought that the soul will survive the decay and dissolution of its earthly investiture. Thus, the late Bishop Butler, whose work is still a standard reference book in many of our leading universities, begins his argument by reference to the metamorphosis of "worms into flies," the hatching of birds from the egg, and even the birth of men from the womb, as so many evidences of a future life; because, he says, "that we are to exist hereafter" . . . is "according to a natural order or appointment of *the very same kind* with that we have already experienced."[1]

Without stopping to show the invalidity of this specific argument (for it has often been refuted), I will proceed with what I have to say regarding the general defects in the system of analogical argumentation when reasoning from purely physical phenomena up to conclusions relating to spiritual laws.

I approach the subject with much diffidence for the reason that this form of reasoning has been resorted to by so many able men that it seems almost iconoclastic to say that it is one of the most unsatisfactory, not to say dangerous, forms of reasoning that can be imagined. Indeed, it is absolutely devoid of the first essential element of correct logical induction.

Analogical reasoning belongs to the realm of poetry and rhetoric, — not to that of logic, nor to that of science,

[1] Butler's Analogy, part i. ch. 1.

except within certain clearly defined limitations. "Poetic license" confers the right to employ almost any figure of speech or comparison, however fanciful; and the same may be said of the productions of the rhetorician. But when we are dealing with scientific questions on a purely logical basis, the field in which analogical reasoning may be properly employed has very decided limitations. It may be proper to employ it when dealing with matters which are known to be governed by the same, or substantially the same, laws; but never when instituting comparisons, either between subjects which are known not to be governed by the same laws, or between subjects which are not known to be governed by the same laws. It seems like arguing a self-evident proposition, to enlarge upon the foregoing; but the necessity for making my meaning clear is evident when we consider the fact that the world has, through countless ages, pinned its faith in a future spiritual life largely upon analogies drawn from the physical universe.

In all inductive reasoning there is one proposition that is, or may be, always assumed; namely, *the constancy of Nature*. Thus, by the observation of a series of phenomena, say the rising and setting of the sun, we are enabled to predict with absolute confidence that it will, on any given day in the future, rise in the east and set in the west. Why? Because we have such confidence in the immutability of the laws of Nature that we assume that the order of the rising and setting of the sun will never be reversed. It is upon this assumption of the constancy of Nature, or rather upon the absolute verity of this assumption, that all advancement in the arts and sciences depends; for if it were not true, we could derive no certain information from our experience or from our observation of the phenomena of Nature. If gravity operated one day and on the next refrained from operating, the whole human race would be instantly put to confusion and lose faith in the integrity of the Creator. Inductive reasoning, therefore, could have no possible value as a means of interpreting the laws of Nature but for the fact that we know that Nature is ever constant.

Reasoning by analogy is one form or modification of induction. It, too, depends for its validity upon the truth of a proposition which is generally assumed to be true. Unlike induction proper, it reasons from the phenomena of one subject up to the general principles pertaining to another subject. It may, or it may not, be a valid form of reasoning; for its validity depends upon the truth of the assumed proposition that *the laws governing the subject-matter observed are identical with those of the subject-matter under investigation.* It is obvious that this proposition must be tacitly assumed, for otherwise there could be no possible excuse for employing that form of reasoning. It is also obvious that if the proposition is true in any given case, the argument is valid; and it is self-evident that if it is not known to be true, the argument is, *ex necessitate*, logically invalid; *a fortiori*, if the proposition is known to be untrue.

Thus, it would be perfectly legitimate for the scientific observer familiar with the natural history of the silkworm to infer the probable metamorphosis of any other larva into a winged insect; because the laws pertaining to the one may legitimately be assumed to be substantially identical with those pertaining to the other. But the case presents a far different aspect when he assumes to reason from the metamorphosis of the caterpillar into the butterfly up to an immortal life for man, for the obvious reasons, first, that the one is an insect and the other is a mammal, so that even the physical laws governing the one are not identical with those pertaining to the other; and, second, that the one retains a physical organization through every change in the metamorphosis, whereas the other is wholly deprived of any bodily organization, so far as our powers of observation inform us, the moment the first change takes place.

Lord Bacon seems to have been fully alive to the intrinsic invalidity of conclusions relating to spiritual life which are drawn from physical phenomena, when he said, —

> "Our inquiries about the nature of the soul must be bound over at last to religion, for otherwise they still lie open to many errors; for, since the substance of the soul was not deduced from the mass of heaven and earth, but immediately from God, *how can the knowledge of the reasonable soul be derived from philosophy?*"

The italics are mine. It is quite certain that if he had lived in a later era he would not have hesitated to set forth, with his accustomed clearness, his utter condemnation of analogical reasoning when employed to demonstrate propositions relating to spiritual law by reference to physical facts. He would certainly have taught mankind the much needed lesson that there is a vast difference between illustration and proof, between poetic license and scientific demonstration.

It seems evident, therefore, that this old and standard argument for a future life must at least fail to be convincing for the very simple and purely logical reason that one of the premises necessary to its completeness is known to be untrue. It does not possess even the negative merit to which the most of Bishop Butler's analogies are limited; namely, that "there is no presumption, from analogy, against the truth" of the proposition advanced. Moreover, "the presumption, from analogy," is decidedly against the continued existence of man after the death of the body, for the obvious reason that the insect dies after the metamorphosis has been completed. Indeed, most of the analogies drawn from our daily observation of the laws of the physical universe lead inevitably to the conclusion that "if a man dies," he does not "live again." For it is a fact within the experience of the most superficial observer that Nature constantly follows the one routine, — birth, growth, maturity, decay, death. Nor does it relieve us of the difficulty to say, as has often been said, that the seed which falls from the tree to the ground contains the same life principle which it derived from the parent stem; that the seed, as a result of its own decay and physical disintegration, springs into renewed life, and another tree is produced, still retaining the same life-principle. Such an analogy can at best be employed to prove only the self-evident truth that a man, in a certain sense, lives in his own posterity. Moreover, the argument is equally as good for pre-existence as it is for future existence. It does not touch the question of the continuance of the individual life after the death of the body; or, if it does, it legitimately leads to the old pagan doctrine of emanation and absorption, which in one form is embodied in the vast system of Buddhism, and in another in that of Averroism. This system supposes that, at the death of an individual, his soul returns to or is absorbed into the universal mind from whom it had originally emanated. Averroes taught the Saracens that the transition of the individual to the universal is instantaneous at death; but the Buddhists maintain that human personality continues in a declining manner for a certain term before nonentity, or Nirvana, is attained.

"Philosophy among the Arabs, and indeed throughout the East, saw an analogy between the gathering of the material of which the body of man consists from the vast store of matter in Nature, and its final restoration to that store, and the emanation of the spirit of man from the universal Intellect, the Divinity, and its final reabsorption.[1]

This is, perhaps, the most plausible analogical reasoning on that subject that has ever been promulgated; but as it

[1] Draper, "Conflict between Religion and Science."

assumes the very thing logically necessary to be proved, namely, that man *has* a soul, and that the soul has a future existence, it must be held not to answer the requirements of logic or of modern science. The doctrine however, with various modifications, is still an essential part of the philosophy of a great proportion of the human race; and Europe itself was only saved to Christianity by the timely establishment of the Inquisition, which carefully eliminated the advocates of the doctrine of emanation and absorption.

In closing my remarks on this branch of the subject, it cannot be too strongly insisted upon that no analogy sought to be instituted between the operations of physical nature and those of the spiritual realm can possess any possible logical validity unless it is first clearly shown that the laws of the two worlds are identical. And as it is manifestly impossible to know the laws which prevail in the unseen universe, it follows that reasoning from such analogies is not only unsatisfactory to the last degree, but, measured by logical and scientific standards, it is, to employ no harsher expression, positively nugatory. It is like trying to demonstrate a proposition in mathematics by citing a rule in grammar. Nor does it avoid the objection to express the analogy in the negative form, which was such a favorite of the late Bishop Butler; for it is the logical equivalent of saying, "There is no presumption, from analogy, to be found in the rules of grammar against the possibility of squaring the circle. Therefore the circle can be squared."

The second in the order of treatment, of the common grounds of the belief in a future life, is *prescriptive authority*. Little need be said on that subject, for the reason that no one, in this enlightened age, claims that the dictum of any man has any legitimate weight as an argument in the absence of facts upon which to base his claims. "Such a doctrine," says Alger, "is the very hiding-place of the power of priestcraft, a vast engine of interest and sway which the shrewd insight of priesthoods has often devised, and the cunning policy of states subsidized. In most cases of this kind the asserted doctrine is placed on the basis of a divine revelation, and must be implicitly received. God proclaims it through his anointed ministers; therefore, to doubt it or logically criticise it is a crime. History bears witness to such a procedure wherever an organized priesthood has flourished, from primeval pagan India to modern papal Rome." No one, of course, holds that the prescriptive authority claimed by the priesthood possessed any scientific value, *per se*, as an argument in favor of a future life; and it is mentioned here only because it is elsewhere set down as one of the grounds of belief in immortality. The basis of the authority of the priesthood is that of divine revelation, which is set down in books which all may read, and each for himself estimate its value as a basis of belief. In the mean time there are few who claim that the Bible records possess any scientific value as arguments in favor of anything therein set forth. There are more who hold modern science in contempt when it sets itself up as a critic of divine revelation; and some go so far as to affect to disdain the principles of induction when they are sought to be applied to the elucidation of the problems of spiritual life. Such men forget that the sole value which any one claims for the records of the New Testament consists in the fact that it is an attempt to prove the doctrine of a future life by the forms of inductive reasoning. What is the New Testament but a record of facts from which the Christian Church proceeds to argue that immortal life for mankind is logically demonstrated? Take, for instance, the record of the life, death, and resurrection of Christ. Here is a fact, or a series of facts, from which the principle of immortality is deduced. Thus Paul pinned his whole faith in immortality on the fact that Christ was raised from the dead; and he used the purest forms of induction to express the grounds of his belief.

"Now, if Christ be preached that he rose from the dead, how say some among you that there is no resurrection of the dead?

"But if there be no resurrection of the dead, then is Christ not risen;

"And if Christ be not risen, then is our preaching vain, and your faith is also vain."[1]

Now, whilst this one fact was a reason all-sufficient to induce Paul to believe in the doctrine of a future life, it does not fulfil the requirements of modern science; not because of any defect in the form of reasoning, but because it is held, first, that the *fact* is not sufficiently authenticated; and, second, that, even if it were perfectly verified, there are other alleged facts which render the conclusion invalid. Thus, it is held that the "twelve men of probity" who are summoned as witnesses of the fact, did not observe the phenomenon under the test conditions required by modern science for the verification of phenomena which are claimed to belong to the domain of the supernatural.

But, supposing the fact of the death and the subsequent resurrection of Jesus to have been verified beyond a scientific doubt, there is another alleged fact which must be considered in that connection. It is alleged that he was a God, equal in power and coexistent with the Father. If that be true, it does not follow, because he had the power to resume his physical investiture after having been crucified, dead, and buried, that a mere man possesses the same power of resurrection after the death of the physical body. In other words, the mere fact that Christ, a God, rose from the dead does not demonstrate the principle of immortality for mankind.

Again, supposing that Jesus was a mere man, invested only with the powers and attributes of common humanity,

[1] 1 Corinthians xv. 12-14.

and hedged about by the same limitations; and that God, as a special manifestation of Divine favor, or for some inscrutable purpose, wrought a miracle in behalf of Jesus and restored him to life, it does not follow that God will repeat the miracle in behalf of each individual for all time to come. If it was a miracle, it was clearly outside of the domain of natural law, and each repetition of it must also transcend the order of Nature. The only other alternative is to suppose that the miracle wrought at the resurrection repealed the old law of Nature and instituted a new one in its stead. We are nowhere taught that a miracle permanently changes the order of Nature. If it did, the miracle at Cana would have changed all the waters of the earth into wine; and the miracle of the loaves and fishes would have released man from that part of the primeval curse which has compelled him to earn his bread by the sweat of his brow.

It will thus be seen that prescriptive authority, even when sanctioned by the words of the only book which has been held by the Christian world to have had a divine origin, is not invested with a sufficient power of conviction to silence the objections of modern science. The sceptical world still demands the same proofs concerning the realities of spiritual life that it requires as the price of its assent to the propositions of material science. I hope, before I close my labors on this volume, measurably to satisfy the demands of intelligent scepticism; but in the mean time I beg the reader to remember that in these preliminary observations I am attempting to give voice to a few of the objections of modern science against the qualitative character of the proofs of a future life afforded by the Bible. My individual estimate of the New Testament records as a proof of immortality will be given in its appropriate place.

The third ground of belief in a future life is the result of *philosophical speculation.* This is a topic of such vast magnitude that it could only be briefly summarized within the limits of a volume like this. It would be foreign to the purpose of this book to undertake such a task, and it could lead to no useful result if all the arguments embraced under this head could be given in full. They all begin and end with the confession of the utter impossibility of demonstrating a future life by scientific methods; or if the author fails to make the acknowledgment, he forces the conviction upon his readers that such is the fact. Thus Alger, in his masterly epitome of the thought of mankind on the destiny of man, from which quotations have already been made, has this confession to make: "The majestic theme of our immortality allures yet baffles us. No fleshly implement of logic or cunning tact of brain can reach the solution. That secret lies in a tissueless realm, whereof no nerve can report beforehand. We must wait a little. Soon we shall grope and guess no more, but grasp and know."

Thus, again, America's greatest philosopher, Emerson, whose sublime faith overreached the bounds of logic and disdained the trammels of science when it failed to reveal what his soul saw so clearly mirrored in the vault of heaven, whilst confessing his inability to give scientific grounds for the faith that was in him, dogmatically asserts that "man is to live hereafter." Continuing, he says: "That the world is for his education is the only sane solution of the enigma." Again, he makes this confession: "I am a better believer, and all serious souls are better believers, in immortality than we can give grounds for. The real evidence is too subtle, or is higher than we can write down in propositions. *We cannot prove our faith by syllogisms.*"

This is the melancholy outcome of all philosophical or metaphysical speculation regarding the destiny of man after the portals of the tomb are passed. It leads us into a maze of doubts and alternate hopes and fears, and ends with the despairing confession that "we cannot prove our faith by syllogisms."

It sounds very unscientific, but I must confess that I attach more of scientific value and importance to Emerson's dogmatic assertion that "man is to live hereafter" than I do to the aggregate of the philosophical speculations known to the literature of the subject. His was one of those pure, lofty, and poetic souls whose intuitive perception and recognition of truth is oftentimes as perfect as a mathematical demonstration. As before remarked, this statement sounds unscientific; but I will endeavor to show, in the proper place, that it is not wholly so. Those who have read "The Law of Psychic Phenomena," especially that part of it relating to the subjective element in poets and poetry, will readily comprehend my meaning.

Perhaps the best specimen of philosophical speculation on the subject is Plato's "Phædo," wherein he puts into the mouth of Socrates an elaborate argument for immortal life. How far it represents the actual opinions of Socrates it is impossible to know, for it is on record that Socrates repudiated some of the sentiments which Plato ascribed to him in some of his earlier works. But as Socrates was dead when the "Phædo" was written, and as Plato was not present on the occasion when the argument was said to have been made, it seems but just to the memory of Socrates to give him the benefit of the doubt. Besides, it is on record that his utterances during his trial do not agree with those ascribed to him by Plato when he was philosophizing with his friends on the day of his death. However, as it is the philosophy, and not the history, of the Greeks that we are discussing, we will relegate that question to its ancient obscurity. The argument, or rather the three arguments, may be briefly summarized as follows: —

His first argument is, that everything in Nature has its contrary. Day follows night, sleep is followed by vigilance, fair is the contrary of foul, justice of injustice, etc. From this he infers that, as life is the contrary of death, it follows that life must succeed death and be produced from it. It might just as well be said that every acid must necessarily become an alkali, or everything bitter must sometime become sweet. It is, in fact, one of those analogical arguments which have been discussed, in which conclusions as to spiritual things are drawn from physical phenomena.

His second argument is based on the assumption that all our present knowledge is merely reminiscence; that is, our acquired knowledge is nothing but the recollection of what we knew in a former state, and that, having existed in a former state, we may confidently count on a continued existence in a future state.

This is a modified form of the doctrine of reincarnation so long held by the Hindu philosophers, and which is now rapidly gaining a foothold in the Western world. It is needless to say that there are no facts to sustain such a doctrine; but the class of minds in which it finds a lodgment cling to it all the more pertinaciously on that account. There are many thousands of people in this country at present who fully agree with Plato in his doctrine of reminiscence; and many of them are full of reminiscences of their own former incarnations. The singular fact of it is that none but the great men of former times appear to have been reincarnated in the nineteenth century. It is also somewhat remarkable that one man can occupy so many different bodies at the same time. I suppose that Socrates at the present moment inhabits some thousands of different modern earthly tabernacles. George Washington is also very generously distributed among the American people. And so of other great men. If we are to believe all that we are told by those who are favored with "reminiscences" of a former life, there are three very obvious deductions which seem inevitable. The first is that no common man is ever reincarnated; second, that the capacity of great men for minute subdivision is illimitable; and third, that reincarnation does not improve the mental capacity of the reincarnated.

It is difficult to treat the doctrine of reincarnation seriously; but from the fact that its followers are becoming numerous it assumes the aspect of a mental phenomenon which must be considered with others of a cognate character. It is, in fact, a psychic phenomenon, and properly belongs to the domain of experimental psychology. The idea originated among a people who for thousands of years have practised hypnotism and kindred arts, and have consequently built up a philosophy upon a basis of subjective hallucinations. Having practised their arts in utter ignorance of the law of suggestion, it follows that their information regarding the other world is just as defective as that obtained in this country through spirit mediums or other forms of hypnotism, and for the same reason. As in spiritistic communications, all that is requisite is the proper suggestion to prove any doctrine whatever; and any one can easily obtain a large and varied assortment of "reminiscences of a former life" by employing a hypnotist and submitting to his manipulations and the proper suggestions. All "reminiscences" of that character may be traced to that or cognate causes.

The third argument of Plato is, that compound substances alone are liable to corruption, or disintegration; and that the soul, being a simple substance, cannot be affected by the death of the body.

This, like the other arguments, is founded on mere assumption without proof. How does any one know that the soul is a simple substance? What facts demonstrate it? Considering the various powers, functions, and affections of the soul, together with the multiplicity of its ideas and emotions, there are just as good reasons for asserting that it is a compound substance as there are for asserting the contrary.

This, then, is the philosophical argument of Plato. It is neither better nor worse than that of any one of his successors who assumes premises that are either not demonstrably true, or are demonstrably untrue. It lacks every essential element of a logical argument; and, were it promulgated to-day for the first time, it would receive the assent of no one acquainted with the elementary principles of correct reasoning. In its day, however, it received the instant and enthusiastic assent of a very large class of people. The doctrine that death is not affliction, but, on the contrary, a direct and sure entrance to a happier life, so influenced the minds of many that they laid violent hands upon themselves in order the sooner to attain that happier life. It is even said that Ptolemæus Philadelphus prohibited Hegisias of Cyrene from teaching it in his school, for fear of depopulating his kingdom. Cicero tells us that it was written of Cleombrotus of Ambracia that, "having paid his last compliment to the sun, he threw himself headlong from the top of a tower into hell; not that he had done anything worthy of death, but had only read Plato's Treatise on the Immortality of the Soul."

It is needless to remark that, as a promoter of suicide, the treatise has long since lost its potency.

The fourth in the series of arguments commonly employed to prove immortality is that of instinctive desire. No more beautiful summary of the argument exists in the English language than that of Addison: —

"Plato, thou reason'st well,
Else whence this pleasing hope, this fond desire,
This longing after immortality?
Or whence this secret dread, and inward horror
Of falling into naught? Why shrinks the soul
Back on herself, and startles at destruction?
'T is the divinity that stirs within us;
'T is heaven itself that points out an hereafter,
And intimates eternity to man." [1]

[1] Addison's "Cato."

"The strongest argument" in favor of immortality, says Cicero, "is that Nature herself is tacitly persuaded of the immortality of the soul; which appears from that great concern, so generally felt by all, for what shall happen after death."

Alger summarizes the argument, and at the same time hints at the answer, as follows: —

> "It is obvious that man is endowed at once with foreknowledge of death, and with a powerful love of life. It is not a love of being here; for he often loathes the scene around him. It is a love of self-possessed existence; a love of his own soul in its central consciousness and bounded royalty. This is the inseparable element of his very entity. Crowned with free-will, walking on the crest of the world, enfeoffed with individual faculties, served by vassal nature with tributes of various joy, he cannot bear the thought of losing himself, or of sliding into the general abyss of matter. His inferior consciousness is permeated with a self-preserving instinct, and shudders at every glimpse of danger or hint of death. The soul, pervaded with a guardian instinct of life, and seeing death's steady approach to destroy the body, necessitates the conception of an escape into another state of existence. Fancy and reason, thus set at work, speedily construct a thousand theories filled with details. Desire first fathers the thought, and then thought woos belief."

As I have before intimated, this is the strongest of all the old arguments in favor of immortality. It is a valid argument as far as it goes, for it is an observable phenomenon — an instinct — of the human mind which points in the direction of a future life. But whilst it is a valid argument, it is not conclusive, for the reason that it lacks the one essential element of a conclusive argument. A phenomenon can only be said to be a conclusive demonstration of the truth of a proposition when there remains no other way of accounting for the phenomenon. This is true, *a fortiori*, when we are seeking to account for a mundane phenomenon by referring it to a supermundane cause. Thus, if man, and man only, desired to live, and if his desire for life had reference only to an existence beyond the grave, and if that desire were clearly shown to be instinctive and universal, then it might be said to be a conclusive argument in support of the hypothesis of a future life. But this "instinctive desire," which so strongly possesses the mind of a man, for a future life, is easily accounted for by reference to that instinct of self-preservation which is proverbially "the first law of nature," is common to all physical organisms, and is no stronger in man than it is in the lowest order of animal life. Man, however, recognizes the fact that his physical organism must perish; but, in the egotism of his manhood, he rebels against the thought of dying as the brute dieth. He looks upon himself as the crowning glory of physical nature. He counts and measures the steps of his evolution from the primordial germ, compares the brief span of his existence with the æons which have been consumed in his production, and concludes that somehow he has been cheated by dissembling Nature of his fair proportion of time and opportunity. At first he rebels against being classed as a lineal descendant of the lower organisms; but the steps of his evolution are too plainly defined in the structure of his predecessors, his pedigree is too clearly written in that of his own, to admit of rational doubt. Compelled to own his relationship to the rest of animated Nature, he finds consolation in the thought that, whilst he may be a product of evolution, he is no longer subject to its laws. He is the product of a process. He is like a machine, which is produced by means of a great variety of processes, but is emancipated from all connection with those employed in its construction the moment it is completed and sent out into the world to perform its functions. Thus, it is argued, is man emancipated from the processes of his evolution and placed upon the apex of Nature, from which point his only means of further progress is by flight into some unknown region where the object of his creation can be accomplished.[1]

With such assumptions does man console himself for his obvious relationship to his fellow worms, and for his lack of time in this life to work out what he fondly conceives to be his mission and destiny. He ignores, or denies, the fact that the same processes of evolution which produced him are still at work in himself and in all his environment, — the same survival of the fittest, though modified by the state of his progress in civilization; the same struggle for life, though modified by the element of an enforced altruism, if such a term is admissible, which compels the inclusion of his race in the object of his struggle. He forgets, too, that the same element which he is pleased to term altruism in himself, is common to many of the lower animals; and that his longing for a future life may be traced to that instinct of self-preservation which he possesses in common with all animated, nay, all organic Nature, and without which the world would soon be depopulated. It seems clear, therefore, that instinctive desire, whilst it is a valid argument as far as it goes, is very far from being conclusive; and must, therefore, for the present, be classed in the same category with many other phenomena of the human mind which seem to point in the direction of a supermundane existence, but logically fail because they are explicable by reference to principles of natural law with which the world is well acquainted.

[1] See Fiske's "Destiny of Man."

CHAPTER III.

SPIRITISM AND HYPNOTISM.

The Phenomena of Spiritism. — Scepticism of the Church. — The Present Attitude of Science. — Spiritistic Phenomena Genuine. — The Two Hypotheses. — The Spirit Medium Self-Hypnotized. — The Intelligence Manifested. — Experimental Hypnotism produces the same Phenomena. — The Power of Telepathy. — The Law of Suggestion. — Suggestion controls the Medium. — The Manufacture of Mediums by Hypnotism. — The Hypothesis of Duality of Mind. — The Objective and Subjective Minds. — The Condition of the Medium and the Hypnotized Subject Identical. — They are governed by the Same Laws. — Socrates as a Roman. — The Spirit of "Cantharides" Invoked. — The Medium not necessarily Dishonest. — The Laws of Telepathy.

I HAVE now briefly reviewed a few of the leading arguments upon which the Christian world has built its hopes of a future life. I have endeavored to show why it is that none of them are convincing to the minds of those who are accustomed to the methods of reasoning which are applied to the solution of the problems of the material universe. It has been shown that no one has attempted to apply the processes of induction to the solution of the great problem, and for the very good reason that, outside of Biblical records, no facts have been adduced, no phenomena have been observed, by the writers on the subject of a future life, upon which immortality for mankind can be legitimately predicated. We now approach a field of observation, however, which bristles with facts and phenomena which millions of our race believe to be demonstrative of a life beyond the grave. It is unnecessary to say that I allude to the phenomena of so-called spiritism.

It has been customary for the Christian Church to ignore the claims of spiritists to recognition as fellow-workers in the realm of spiritual philosophy. It has derided their pretensions to an experimental knowledge of the truth of one of the essential doctrines of the Christian Church; namely, the doctrine of immortal life. It has persistently denied the genuineness of their phenomena; or, where compelled to admit the verity of the manifestations (which differ in no essential particular from those recorded in Holy Writ), it has attributed them to diabolical agency. Scientists, until within a very few years, have been content with a general denial of the existence of the phenomena, and a disdainful refusal to investigate. Their attitude is identical with that of one of their number, who, when called upon to explain the phenomenon of the fall of meteoric stones, exclaimed: "There are no stones in the air; therefore no stones fall from the air." The materialistic scientist says, "There are no spirits; therefore there are no spiritistic phenomena."

Happily for mankind, and much to the credit of a vast number of consistent members of the Christian Church, as well as of thousands of the ablest scientists in the civilized world, this attitude is no longer popular, but is fast giving way to one of intelligent and honest investigation. This change is largely due to the London Society for Psychical Research, which comprises among its members a large number of scientists whose reputation as careful investigators in the realm of natural science is international. The result is that there is no longer a rational doubt of the genuineness of so-called spiritistic phenomena among those who have taken the trouble to apply the strict rules of scientific inquiry to the subject-matter. They declare that no phenomenon in the realm of physical science is better authenticated than those of so-called spiritism. This being true, it follows that the causes must be investigated with the same care and in the same spirit of candor that has characterized the investigation of the fact; and to that end the millions of human beings who have claimed a supermundane origin for the phenomena are entitled to a respectful hearing. For, if their hypothesis is demonstrably true, the question of spirit life is no longer a speculative problem; and if it is not true, it is important that the world should know to what power or law of Nature the phenomena are to be attributed.

In dealing with the phenomena under consideration I do not propose to waste the time of the reader by the discussion of each particular phase of manifestation. A volume of the size of this would be all too small to discuss exhaustively the many-sided problem, or to explain the various characteristics of the phenomena. Besides, it would be but a repetition of what I have already done in another work.[1] It will only be necessary here to discuss the one salient feature which is common to all the phenomena; and that is the intelligence which is manifested. This intelligence claims to be from the denizens of another world; and spiritists hold that there is indubitable evidence in the manifestations themselves that they proceed from spirits of the dead.

It will, however, be necessary to discuss the subject of spiritism at some length, for the reason that in the consideration of scientific problems it is of the very first importance that the phenomena under consideration should be properly classified. There are two hypotheses employed to account for so-called spiritistic phenomena. One is that they proceed from disembodied spirits; and the other is that they are produced by the conscious or unconscious

[1] For a full discussion of the various phases and characteristics of spiritism and other psychic phenomena, see the author's work entitled "The Law of Psychic Phenomena."

exercise of powers inherent in the living man; and that the known powers of the embodied human soul are sufficient to account for all that is mysterious in the phenomena. Both these hypotheses cannot be true. One must be true and the other false. There can be no compromise, as some spiritists would have us believe. They are not concurrent hypotheses. They are absolutely antagonistic. Hence the importance of properly classifying the phenomena at the threshold of our argument. For if it is found that they are produced by the living, then we have a solid basis of fact from which we can deduce the most momentous conclusions regarding the destiny of man. But if it can be demonstrated that the whole, or any part, of spiritistic phenomena are produced by disembodied spirits, the whole subject is thrown into logical chaos; for *a future life for man is not demonstrated by showing that spirits communicate with the living*, — for the very obvious reason that we still have no means of determining whether any communicating spirit is that of one who has once lived upon the earth, or is an "evil spirit," or an "elemental," or an "elementary," or a "devil," or any other of the denizens of the other world with which it has been peopled by superstition.

It is well known to all observers of spiritistic phenomena that the one essential prerequisite to their production is the presence of a so-called "medium." The term "medium" has been bestowed upon those who are instrumental in the production of the phenomena, because of the assumption that the intelligence conveyed is from spirits of the dead to the living. The instrument through which these messages are conveyed is, therefore, designated as a "medium." This term, consequently, implies a theory of causation; and as it is better to avoid giving an implied assent to any theory by the employment of a careless terminology, I shall hereinafter employ the more non-committal term of *psychic* to designate the person in whose presence these manifestations occur.

The psychic is usually seated at a table around which several others are gathered, and the collective company is designated as a "circle." After all are seated and quiet is secured, the psychic enters into a state which may be described by the generic term of trance. This state, it may be premised, is identical with that of hypnosis, and it varies in depth from that of an apparently normal condition to that of profound objective insensibility. This condition is self-induced, and during its continuance various phenomena are produced; though each psychic is usually confined to one class of manifestations. In the presence of some psychics percussive sounds are heard, called "spirit raps." These are sometimes heard on the table, at other times on the floor, or on the walls, ceiling, or furniture of the room. In presence of other psychics the table is tilted, and oftentimes it levitates into the air without physical contact with any one. Some psychics write automatically; that is, they seize a pencil and write, their hand seemingly being moved by an extraneous force which acts independently of the conscious volition of the psychic.

It would be tedious even to enumerate the different forms which these manifestations assume, and it would be foreign to my purpose to do so; but there is one characteristic common to all the phenomena. They all manifest intelligence, and this intelligence is almost invariably exercised independently of the conscious volition of the psychic. If the psychic is what is known as a "writing medium," messages will be written purporting to emanate from spirits of the dead; and the information conveyed will often transcend the conscious knowledge of the psychic, and sometimes the messages will convey information not in the conscious possession either of the psychic or of any one else in the room. So perfect is the automatism of some psychics that they will write long messages, characterized by more than ordinary intelligence and by perfect coherency, and at the same time carry on an animated conversation with others in the room, and on a subject entirely foreign to that of the message which they are writing. Others have been known to write normally on one subject with the right hand, while at the same time the left was automatically writing a message on another subject, the latter purporting to emanate from the spirit of some one who is dead.

In fact, the intellectual feats performed by some psychics almost transcend belief, and, were they not abundantly authenticated by the most severe scientific tests, would be unworthy of credence. As it is, they demand investigation by the strictest rules of logical induction, with the view of testing the validity of the hypothesis that they are of supermundane origin. In doing so we should always bear in mind the fundamental axiom of science that we have neither occasion nor logical right to attribute any phenomenon to supermundane agency if it is explicable by reference to natural laws. On the other hand, if they cannot be thus explained, those who hold to the supermundane explanation have a right to demand that their hypothesis shall be provisionally accepted. A fact is a fact, and a psychic fact is just as much a fact as a waxing and waning moon. Science has no more right to ignore the one than the other. No fact in Nature can safely be ignored, for no fact is wholly insignificant. Each is so inseparably bound up with the others that if even the most apparently unimportant fact is left out of consideration, inextricable confusion is likely to follow. An apparently insignificant fact is like a cipher in arithmetic. Separately considered, it has no value; but in its proper place it increases the value of the other figures tenfold. Drop one out at any given point, and the whole calculation results in a wrong conclusion. This is why the old psychology is incomplete, confused, and unsatisfactory. It was built up on a basis of speculative philosophy, and, necessarily, in utter ignorance of a vast array of psychic facts and principles which have transpired and been discovered within the last quarter of the nineteenth century.

It is the same with speculative theology. Never, before the development of experimental psychology, has it been possible to demonstrate, with anything like scientific accuracy, the fact that man has a soul. With the advent, however, of hypnotism, mesmerism, and cognate psychic phenomena, all has been changed; and it is now possible to demonstrate the truth of many propositions that have heretofore lain wholly in the realm of speculative philosophy, or have been relegated by the scientific world to the domain of superstition. New discoveries are constantly being made in the psychic world, — discoveries which promise soon to place psychology fairly within the domain of the exact sciences. Then will man, in the truest sense, be enabled to "look through Nature up to Nature's God," not by means of vain analogies drawn from the realm of gross material existence, but by studying his own powers and attributes. The divine pedigree of man and his title-deed to immortal life are written upon the tablets of his own soul, and not upon the wings of the Lepidoptera.

Experimental hypnotism demonstrates several important characteristics of man's mental organization which throw a flood of light upon many obscure problems of psychology. The first and most important of these characteristics is the fact that man possesses a dual mental organization. This is an old doctrine which has been held by speculative philosophers from Plato down to the present day; but it was never scientifically demonstrated until hypnotism revealed it as a law of the human intellect. When a person is perfectly hypnotized, his objective senses are put to sleep. He can neither hear, see, smell, taste, nor feel, except at the will of the hypnotist who induces the lethargy. At the bidding of the latter, however, the subject can be roused to a state of intense activity and power. His mind then seems to be completely transformed, and to possess powers and attributes which were entirely foreign to the subject in his normal condition. His memory is exalted, and in many cases it seems to be practically perfect. In many instances he develops the power of telepathy, — that is, the power to read the mind of the operator, or of any one with whom he is *en rapport.* This is one of the most important things to be remembered in this connection. The power of reading the minds of those with whom the hypnotized person comes in mental contact is the master key which unlocks many of the grand mysteries of psychic phenomena. That power, in connection with the perfection of his memory, constitutes a salient feature of the accomplishments of every well-developed psychic.

There is, however, another characteristic of the psychic's mind when in the hypnotic or partially hypnotic state, which constitutes the greatest and most important discovery of modern experimental psychology. It is this: when hypnotized, the subject is constantly amenable to control by suggestion; that is to say, he accepts as absolutely true every statement that is made to him. Thus, if he is told that he is the President of the United States, he will immediately accept the statement as true, and assume all the airs of importance and dignity that he may conceive to be the legitimate concomitant of that more or less exalted position. If he is then told that he is a street mendicant, he will immediately change his demeanor and assume an attitude of humble suppliancy. In short, he may be made to believe that he is anything, animate or inanimate, and he will act the part suggested with wonderful fidelity to nature, just so far as his knowledge of the characteristics of the person or thing suggested extends. Thus, if it is suggested that he is some one of his intimate acquaintances, he will immediately proceed to imitate all the salient peculiarities of his friend in voice, tone, gesture, and favorite topics of conversation.

Again, it may be suggested to him that he is the spirit of some deceased friend or acquaintance. It matters not. He will confidently believe the suggestion to be literally true, and will assume the characteristics of the deceased, and will, if interrogated, give a full account of his surroundings in the spirit world, albeit his account of his spirit abode will be in exact accordance with his own preconceived ideas on that subject. In other words, the suggestions embraced in his education will give character to his account of his imaginary spirit environment. These facts are well known to all hypnotists; and any one familiar with the works of the mesmerists of the first half of the present century will recall a thousand instances illustrating what has been said. Those works were written, and the experiments made, in utter ignorance of the inexorable law of suggestion; and, hence, many believed that the mesmerized subjects were actually in communication with the spirit world. One man[1] devoted his life to hypnotizing people, sending them to the spirit land, and recording their accounts of what they saw. It is unnecessary to remark that his book was for many years a standard authority among spiritists, — a book of reference, by consulting which all disputed questions relating to the topography, sociology, or climatology of the spirit world could be definitely settled.

Again, it is well known to hypnotists that a suggestion to a subject that he is under the control of a spirit will result in the production of all the phenomena of spirit mediumship, limited only by his lack of training as a psychic. It is also well known that the quickest and surest way to train a psychic for spirit mediumship is to hypnotize him often enough to produce the requisite neurotic condition, accom-

[1] Cahagnet.

panying each hypnotization with suggestions of spirit presence and spirit control. Other things being equal, the best psychics are those who have been developed by hypnotic processes. It is perfectly easy by suggestion to train a psychic to habits of self-hypnotization; and when that is accomplished, he is ready to enter the field as a full-fledged medium of communication between the two worlds.

It is impossible within the space at my command to give even a *résumé* of the many characteristics of the hypnotized subject which go to establish the fact of duality of mind. Nor does it matter, for the purposes of this discussion, whether we regard man as being possessed of two minds, each possessing independent powers and attributes, or regard his one mind as being possessed of certain diverse powers which manifest themselves differently under varying conditions. The fact remains, however, that everything happens just as though man were possessed of a dual mind; and we have a logical right, therefore, to assume it to be true as a provisional hypothesis. Besides, having fully discussed that question elsewhere,[1] I cannot, without unseemly repetition, discuss it exhaustively here.

I have assumed, therefore, that man possesses a dual mind. For the sake of clearness, as well as for the want of a better term, I have designated one as the "objective mind" and the other as the "subjective mind."

"The objective mind takes cognizance of the objective world. Its media of observation are the five physical senses.

"The subjective mind takes cognizance of its environment by means independent of the physical senses. It perceives by intuition. It is the seat of the emotions, and the storehouse of memory. It performs its highest functions when the objective senses are in abeyance. In a word, it is that intelligence which makes itself manifest in a hypnotic subject when he is in a state of somnambulism."[2]

[1] See "The Law of Psychic Phenomena."

[2] Op. cit. p. 29.

In adopting this terminology I have merely followed that of Averroes, as given by Professor Draper in his "Conflict between Religion and Science;" although the meaning which he attaches to his term — "objective intellect" — differs materially from that of my definition of the objective mind. His term — "subjective intellect" — is explained as follows: —

"The individual, or passive, or subjective intellect is an emanation from the universal, and constitutes what is termed the soul of man."

This perfectly expresses my belief regarding the subjective mind. It not only possesses powers and functions which act independently of those of the objective mind, but its very manifestation shows it to be a distinct entity, and apparently capable of maintaining an existence independently of the body. It is a spark of the Divine Intelligence. It is the soul.

I have now sufficiently enlarged upon the subject of the dual hypothesis to enable the intelligent reader to grasp my meaning. The law of suggestion has also been clearly explained. It has also been shown that telepathy, or mind-reading, is a power of the hypnotized subject. The three propositions of my hypothesis, therefore, stand thus: —

1. Man is possessed of a dual mind, — objective and subjective.

2. The subjective mind is constantly amenable to control by suggestion.

3. Telepathy is a power of the subjective mind.

With these three fundamental propositions clearly before us, we are prepared intelligently to compare the phenomena of spiritism with those of hypnotism, with the view of a candid inquiry whether there is any phenomenon produced by the one that cannot be reproduced by the other. Or, to put it in another form, is there anything in the phenomena of spiritism that cannot be explained by reference to the known powers inherent in the living man as developed by and through the science of hypnotism?

A brief comparison of the two classes of phenomena will make a *prima facie* case against the spiritistic hypothesis. It would, however, be more in accordance with the principles of logical scientific investigation to say that the spiritistic hypothesis is, *prima facie*, untrue, and that the *onus probandi* is upon those who claim a supermundane origin for any phenomenon whatever. But we will waive our logical rights for the moment, and proceed to assume the burden of proof.

In the first place, the condition of the psychic when producing spiritistic phenomena is identical with that of the hypnotized subject. The only difference between the two is not in the condition, but in the method of inducing the condition. Both are hypnotized; but the psychic is self-hypnotized by an auto-suggestion, whereas the hypnotic subject is hypnotized by the suggestion of another. It is well known, however, that any hypnotic subject can easily be trained to hypnotize himself. When that is done, all the conditions requisite to successful "mediumship" are present in the hypnotic subject; and if he believes in spiritism, the suggestion embraced in that belief will do the rest. It may be objected that the spiritistic psychic often produces his phenomena while apparently in his normal condition. To this it is answered that there are also an infinite number of degrees of hypnotism which shade into each other imperceptibly, ranging from the apparently normal state to that of profound hypnotic lethargy. Thus, Bernheim[1] was enabled to produce "all suggestive phenomena up to hallucination," while the patient was in the "waking condition;" that is, in a condition that could not be distinguished from the normal by any ordinary tests. That the subject

[1] Suggestive Therapeutics, pp. 79, 81, 83.

was actually hypnotized, however, was demonstrated by the very fact that he was controllable by suggestion. In one case he produced such a perfect state of analgesia, by mere suggestion, the patient being in the "waking condition," that the application of Dubois Raymond's electrical apparatus, with "the greatest current attainable" turned on, produced no sensation whatever; although, as Bernheim remarks, "the painful sensation thus produced is normally unbearable."

It will thus be seen that the degree of hypnosis has no necessary effect upon the manifestations, either in hypnotism proper, or in the psychic phenomena of spiritism. Practice in each case seems to develop the suggestibility of the psychic and his consequent ability to produce the various phenomena while in a condition apparently closely approaching the normal.

That the condition of the psychic and that of the hypnotic subject are identical, is further demonstrated by the fact that they are governed by the same laws. The most important of these laws is that of suggestion; and all the facts of spiritistic phenomena show that the psychic is constantly dominated by that subtle power. His very entrance into the psychic state is produced by the suggestion embraced in his belief that he is about to pass under the control of an extraneous force which he believes to be a spirit. When he is thus self-hypnotized, he is necessarily amenable to the same power. This is clearly shown by the well-known fact that any spirit can then be invoked by those present, and one spirit will respond just as readily as another. Besides, the spirit of a living man will respond with as much alacrity as that of a dead man, provided the question is asked in such a way as to cause the psychic to believe that the bearer of the name is dead. Moreover, the spirit of a purely imaginary person is just as responsive to an invocation as any other. The writer once attended a spiritistic séance presided over by an ignorant psychic. Some one asked for the spirit of Socrates; and the old philosopher promptly responded. His communication was couched in terms that were evidently intended to be somewhat lofty, and were so considered by the enthusiastic admirers of the psychic; nor was their admiration at all diminished by the fact that Socrates seemed to labor under the impression that he was a Roman when on earth. This was afterwards explained by a prominent local authority in spirit philosophy by saying that those old spirits had reached an altitude so far removed from earthly life that they were no longer interested in mundane affairs, and many of them had really forgotten their earthly names and nationality. A wag who was present asked for a communication from "the ancient Greek philosopher, Cantharides." This request was also promptly complied with in an equally lofty strain of bad English. When asked for a description of the latter personage, the psychic described an old man with long white hair, a flowing beard, and a "very high forehead." This goes to show either that the psychic was dominated by an absurdly false suggestion, or that evolution is more rapid in the spirit land than it is here; for the description of the personality of "Cantharides" certainly did not suggest a coleopterous ancestry.

To do the psychic entire justice, let me say that a circumstance like the foregoing does not, in the remotest degree, impugn his honesty or sincerity. He is in a hypnotic condition. Being in that state, he is necessarily dominated by the laws pertaining to it. His normal reason is in abeyance. His subjective mind is active; and the one all-potent, never-failing law of subjective mental activity is the law of suggestion. Like every other law of Nature, there are no exceptions to its inexorable rules. He believes, because he must believe, every suggestion made to him. The suggestion enforced by the current theory of spiritism convinces him that he is a chosen medium of spirit communication with the inhabitants of this world. His reason confirms the belief; for he finds himself dominated by what he believes to be an extraneous force which seems to act independently of his conscious volition. This force is found to possess a remarkable intelligence. It will answer questions, and write essays, poems, and polemics with equal facility, and it often imparts knowledge and divulges secrets of which he is not the conscious custodian. It gives information, which he knows he never could have possessed, concerning the affairs of his auditors, — secrets, perhaps, which the latter declare could never have been known to any but themselves and some deceased friend. When the intelligence is interrogated, it declares itself to be the disembodied spirit of the friend who was the joint custodian of the sitter's secret.

With all this array of evidence before him, not only of spirit presence but of spirit identity, what is his natural conclusion? He is not a scientist, and does not, therefore, know that it is unscientific to believe that man has a soul. He was taught at his mother's knee that he not only has a soul, but that it is destined to an immortal existence. He has never heard of the dual nature of the mind of man, he knows nothing of the law of suggestion, and "telepathy" is not in his vocabulary. But he has common sense, and he is not aware that it is unscientific to exercise it when dealing with phenomena which he cannot explain otherwise than as being tangible evidence of the truth of what he has always been taught to believe was the essence of the teachings of Holy Writ. And he does believe it, honestly and implicitly. It is henceforth his religion, his consolation in this life, and the sheet anchor of his faith in immortality. With all the evidence before him, and in the absence of any other rational explanation, he would be an unreasoning sceptic if he did not believe it.

It will thus be seen what an all-potent suggestion dominates his subjective mind. By virtue of the fundamental law of its being it must accept every suggestion imparted to it, however absurd or contrary to objective knowledge and experience. But when it is confirmed by objective reason and reinforced by the tenderest emotions and loftiest aspirations of the human soul, it becomes a dominant power which cannot be resisted. In this state of mind, objective and subjective, the suggestion that a spirit from another world is in possession of the psychic's hand and guiding its movements, is, and must be, seized upon by his subjective mind and implicitly believed and acted upon, and the suggested spirit personated with all the preternatural acumen and dramatic circumstance characteristic of subjective mental activity. If it were not so, then there would be a break in the operations of a law of Nature, — an exception to the universal rule, which in itself would constitute a miracle as great as would be the suspension for a day of the law of gravitation.

Another fact which correlates the phenomena produced by the spiritistic psychic with those of hypnotism is that the psychic in each case develops the power of telepathy, or mind-reading. I shall not waste time in this connection by offering proofs of the reality of this power. That work has been most thoroughly done by the Society for Psychical Research. It is sufficient to say at this time that no law or power of Nature has been more completely and scientifically demonstrated than has been the law of telepathy. There is, however, one important principle pertaining to the subject-matter which must be understood before the full significance of the fact relating to it can be appreciated or comprehended. *Telepathy is the means of communion between subjective minds.* The objective mind does not necessarily participate in the communication. The message, in other words, is not necessarily on a subject of which either party is consciously thinking. It frequently has no connection whatever with the conscious thoughts of either of the participants in the production of the phenomena. A message of the utmost importance may, therefore, never rise above the threshold of the "percipient's"[1] consciousness or even be consciously sent by the "agent." It requires some degree of psychic development to enable one to become conscious of the reception of a telepathic message. Hence it is that comparatively few are able to perceive the details of a communication; although there are few who have not felt an unaccountable impression which is afterwards discovered to have a telepathic origin, or is classed as a "coincidence" by the sceptical. It requires an extraordinary, even an abnormal development of psychic power to enable one consciously to read the thoughts of another in detail. That power is possessed by few outside the circle of so-called spirit mediums, and of those who have developed it by hypnotic processes for purposes of public exhibition.

It is thought that enough has been said to correlate the phenomena of spiritism with those of hypnotism, at least so far as the conditions necessary for the production of the two classes of phenomena are concerned. In the next chapter the principles herein set forth will be applied to the elucidation of the phenomena alleged to be produced by disembodied spirits. I will also take occasion to answer some objections urged by scientific students of spiritism who hold that there is still a "small residuum" of phenomena which is not explicable by reference to known laws of Nature.

[1] The "percipient," in the vocabulary of psychic science, is the one to whom a message is sent, or who witnesses a phenomenon. The "agent" is the one who sends the message, or produces the phenomenon.

CHAPTER IV.

SPIRITISTIC PHENOMENA.

The Typical Séance. — "Test" Cases. — The Way Proselytes are made. — The Telepathic Explanation. — What Telepathy is. — Views of Rev. Minot J. Savage and of Mr. F. W. H. Myers. — Their Test Cases Explained. — The Small Residuum of Phenomena which they cannot account for. — The Shipwreck. — An Alleged Spirit Communication from a Victim. — A Telepathic Explanation. — Telepathy *vs.* Clairvoyance. — A Typical Case. — "Stretching" the Theory of Telepathy. — Views of Mr. Podmore.

THE following propositions have now been provisionally established: —

1. The condition of the spiritistic psychic or "medium" and that of the hypnotic subject are identical at the time when their respective phenomena are produced.

2. That condition in both is what is known to science as hypnosis, or partial hypnosis, as the case may be.

These propositions are demonstrated by the following facts: —

1. The psychics in each case are in a more or less profound state of objective insensibility.

2. That state or condition in each case is induced by suggestion.

3. In each case the psychic is constantly amenable to control by suggestion.

4. In each case the power to read the minds of others is developed by persistence in the practice of entering the hypnotic or subjective condition.

It has also been established, as a corollary of the foregoing propositions, that in the mind of each psychic a subjective hallucination can be induced by a suggestion; and that it is a matter of indifference whether it be an autosuggestion, a telepathic suggestion, or an oral suggestion.

It is unnecessary to consume much time in the application of these facts to the ordinary manifestations of that intelligence which is alleged to proceed from disembodied spirits. It may not be amiss, however, to present an imaginary case exactly corresponding to those of every-day experience. The *dramatis personæ* are a well-developed psychic and a client who is seeking for tangible evidence of a future life, and will be satisfied with nothing short of a "test case" of spirit identity. He is full of hope that he may be put in possession of indubitable evidence of the reality of spirit intercourse with the living, for he longs to establish communication with the loved and lost, — longs "for the touch of a vanished hand, and the sound of a voice that is still." But he does not mean to allow his emotions to warp his judgment, or to entertain a belief whose parentage can be traced to his desires alone. Perhaps he has heard of mind-reading, and feels prepared to detect any evidence of the exercise of that power, or of any species of mental legerdemain. He takes care in the selection of a psychic, and seeks one who is utterly ignorant of his name, local habitation and antecedents. Having found one possessing all the necessary qualifications, he seats himself and awaits results.

The psychic enters the subjective state, and presently begins his revelation. He begins by making an exhaustive inventory of his client's mental qualifications, whom he asserts to be a man of exalted character and gigantic intellect. He tells him that all he needs is opportunity to make his mark in the world; that he is not as well off in this world's goods as he would like to be, but that fame and fortune are near at hand; that he is a fearless investigator, but is not easily fooled, etc., etc. All this, however, in the estimation of the client, proves nothing but the good judgment of the psychic and the clearness of his perception of human character. Anybody might know all this at a glance, for that matter. It is not even good evidence of mind-reading.

Presently the psychic tells the client's name. He is somewhat startled, but reflects that it may have been obtained from the hotel register, or that it may be mind-reading. The psychic then describes a spirit of which he sees a vision, relates all the circumstances of the death and burial of the person, and, perhaps, states the name of the deceased, the relationship borne to the client, and many little details which may or may not be true. In its important features, however, the statement is exact. It is the very person the client most desires to communicate with; but he is not satisfied. All this is plainly within the domain of mind-reading, and he is not to be deceived. But he is interested and hopeful, and asks for a communication, which is given. It is couched in endearing, but general terms; plenty of good advice is given, and it ends in a rhapsodical expression of assurance of a life beyond the grave and of a happy reunion when life's fitful fever is over. The client is much affected, but reflects that all this is plainly within the capacity of the psychic. What he is after is a "test," and he so informs the psychic. Presently the "spirit" relates some little episode which the client knows was within the exclusive knowledge of himself and the deceased. His doubts begin to vanish; but he reflects that the knowledge of the circumstance was in his mind, although he does not remember of having thought of it that day. The test is not conclusive, and he awaits further developments.

Up to this point the spiritist of average intelligence will agree that no valid evidence of spirit identity has been forthcoming. It is obviously all within the domain and possibilities of telepathy.

Presently another "spirit" appears upon the scene. The psychic describes it with great minuteness, and finally gives its name. The client is confounded. "This cannot be mind-reading," he soliloquizes, "for I have not thought of that man for twenty years." He has crossed the Rubicon. The limit of his knowledge of telepathy has been reached, and to him it is no longer a tenable hypothesis. As if to make assurance doubly sure, the "spirit" recalls a business transaction between himself and the client which took place forty years agone. The client racks his memory in vain for a trace of recollection of the transaction. He does not even remember of ever having had any business with the deceased. He knows, however, that he has the means of verifying the statement if it is true, and he closes the séance and hurries home to institute a search of his old records. He finds, to his surprise and delight, that his spirit friend is right in every particular.

A proselyte is gained for spiritism. Henceforth he haunts spirit mediums, and spirit mediums haunt him. Perchance his mind is not so warped that he ceases to recognize telepathy as a possible factor in alleged spirit intercourse. He may even admit that the great bulk of spiritistic phenomena can be accounted for by reference to telepathy; but he holds that there is a "small residuum" of phenomena that cannot be thus explained.

Until within a very few years it would have been impossible, by invoking the aid of telepathy, to account for the phenomena which converted our friend. But it has now been ascertained that even if one does not happen to be consciously thinking of a particular spirit when that spirit is announced and correctly described, such a mental condition does not militate against the telepathic theory. On the contrary, it is clear that what one is consciously thinking of has no necessary connection with the subject of a telepathic message. Telepathy, as we have already seen, is the means of communication between sub-conscious intelligences; and it is only those who are psychically developed who can become conscious of the operations of the subjective intelligence. There are several things to be considered in this connection: —

First, it must be remembered that it is the subjective mind that reads, and is read, telepathically.

Secondly, the memory of the subjective mind is prodigious and practically perfect; and it often happens that circumstances entirely forgotten by the objective mind of one who consults a psychic are recalled by his subjective mind. Hence it is that when the client declares that the circumstance never occurred, and afterwards ascertains that he was mistaken, it is no evidence whatever of the agency of disembodied spirits. Moreover, when we consider the perfection of subjective memory and the imperfection of objective recollection, it is extremely hazardous for any one to say positively that he never knew a particular circumstance or thing. Such a statement can only be accepted as evidence that he does not recall the fact stated.

Thirdly, the law of suggestion is an important factor which must be considered in the solution of the mystery. It is obvious that when two subjective intelligences are in communication, and each is dominated by the suggestions of spiritism, and one is anxious to convince the objective intelligence of the other of the reality of spirit intercourse, and the other is anxious to be convinced and is seeking for a test case, the best tests within the combined resources of the two intelligences are more than likely to be forthcoming. This is certainly borne out by the experience of every intelligent investigator of the phenomena of spiritism. I should not consider myself justified in supposing a subjective conspiracy to be possible, even under the inexorable law of suggestion, if a case of spirit identity had ever been made that was free from the doubts thrust upon us by the known laws of telepathy and suggestion.

Thus far I have said nothing the essential features of which scientific investigators of the phenomena of spiritism will not indorse. Indeed, I believe that intelligent spiritists very generally concede that phenomena of the character outlined in the foregoing remarks are not free from doubt; and they are ready to accept substantially the explanation here given. I now approach a branch of the subject, however, about which the most scientific investigators of the phenomena are at variance. It has been, and is now, my desire and purpose to avoid controversial argument on these topics; but I feel impelled by my very respect for two gentlemen who have arrived at slightly different conclusions from mine, to place my views alongside of theirs, so to speak, in order to allow our readers to judge which presents the better reason for the faith that is in him. I do so for the further reason that my desire to do exact justice to the cause of spiritism impels me to state the position of two of its ablest champions with whose views I am acquainted. They are gentlemen whose reputation for learning and scientific attainments and for candor and transparent honesty of purpose is as wide as civilization, and places them at the head of the list of scientific champions of the cause of spiritism. It is almost superfluous to now say that I refer to Rev. Minot J. Savage, of Boston, and F. W. H. Myers, of London. Neither of them needs any introduction to the English-speaking world. To Mr. Myers the world owes more than it can ever repay for his indefatigable and purely scientific labors as Secretary of the London Society for Psychical Research. To Mr. Savage the world owes much for his championship of free thought, and of progress in every field of human endeavor.

If I understand the views of Mr. Savage and Mr. Myers,

they agree in their interpretation of such phenomena as I have outlined, and each would offer substantially the same explanation that I have suggested. But at this point our paths diverge. Each of those gentlemen is willing to admit that when a psychic transmits a message to his client containing information which is in his (the psychic's) possession, it cannot reasonably be attributed to the agency of disembodied spirits. Each is also willing to admit that when the message contains facts known to some one in his immediate presence and with whom he is *en rapport*, the agency of spirits of the dead cannot be presumed. Each will doubtless admit that sub-conscious memory may enter as a factor in the case, and that the sub-conscious intelligence — or, to use the favorite terminology employed by Mr. Myers to designate the subjective mind, the "subliminal consciousness" — of the psychic or that of his client may retain and use facts which the conscious, or objective, mind may have entirely forgotten. All these considerations, and many others, have doubtless entered into the calculations of both those gentlemen when estimating, in their careful and severely scientific way, the weight of evidence for and against the hypothesis of spirit intercourse. But here is where they pause. They are not willing to admit that telepathy can possibly enter as a factor in the case when the message contains facts not known either to the psychic or his client or any of the psychic's friends. I will now let Mr. Savage speak for himself. In an able address[1] delivered before the American Psychical Society, he used the following words: —

"In the presence of psychics I have been told things which I know the psychic did not know and never had known. There is no longer the least shadow of a doubt of that in my mind. But I have always said this does not go far enough, possibly this may mean telepathy only. Although the psychic is not a

[1] Psychical Review, vol. i. no. 1.

clairvoyant, is not conscious of possessing any means of getting at the contents of my mind, yet the psychic's mind may be a mirror in which my thoughts and knowledge are reflected, and I may be getting back only what I have given. So when I have gone only to that extent I have felt that I have not gone far enough to convince me that I was dealing with a disembodied intelligence. But under certain conditions I have gone farther than that; and it is these other cases that we are always looking for as additional proof, — these cases in which I have received communication concerning something which neither the psychic nor myself knew. There have been several cases not only in my own experience, but, more still, in the experience of persons whose judgment and power of investigation I trust as I trust my own, in which there has been the communication of intelligence that neither the psychic nor the sitter possessed nor ever did possess. I have had it in such circumstances as this. I have had communication while sitting in my study concerning things that were taking place two hundred miles away. Over and over again occurrences like this have taken place, and I submit that my knowledge of science and philosophy does not give me any hint of an explanation for these things. It seems to me to be stretching the theory of telepathy and clairvoyance beyond probability to call them in to explain them. I do not know what to make of them except on the theory that some third and invisible intelligence was concerned."

In a short but brilliant essay[1] entitled "Science and a Future Life," Mr. Myers thus sums up his own views, and locates the line which marks the boundary between the two worlds: —

"I am further strengthened in this belief by the study of the automatic phenomena briefly noticed above. I observe that in all the varieties of automatic action — of which automatic writing may be taken as a prominent type — the contents of the messages given seem to be derived from three sources. First of all comes the automatist's own mind. From that the vast bulk of the messages are undoubtedly drawn, even when they refer to matters which the automatist once knew, but has en-

[1] This article is printed in a volume with several other valuable essays on cognate subjects by the same author.

tirely forgotten. Whatever has gone into the mind may come out of the mind; although this automatism may be the only way of getting at it. Secondly, there is a small percentage of messages apparently telepathic, — containing, that is to say, facts probably unknown to the automatist, but known to some living person in his company, or connected with him. But, thirdly, there is a still smaller residuum of messages which I cannot thus explain, — messages which contain facts apparently not known to the automatist nor to any living friend of his, but known to some deceased person, perhaps to a total stranger to the living man whose hand is writing. I cannot avoid the conviction that in some way — however dreamlike and indirect — it is the departed personality which originates such messages as these."

Here we have the clear and concise statements of two men, representatives, respectively, of the best thought, on that side of the question, of two hemispheres, each defining the same limit of scientific explanation, on mundane principles, of spiritistic phenomena. These gentlemen, being, *par excellence*, the representatives of scientific spiritism on the two sides of the Atlantic, are entitled to a full hearing and a candid and respectful consideration. Their views, however, as quoted above, are stated in abstract form, and their exact meaning may not be readily comprehended. In order, therefore, that the reader may have a concrete illustration of just what is meant, it will be necessary to quote further from their works. Fortunately for our purpose, Mr. Savage has issued a volume [1] recently which is filled with well-authenticated facts which the author of the book cannot explain by reference to telepathy. I select at random the following case, which is fairly illustrative of the "sticking point" in its most pronounced, not to say virulent, form. In order to do Mr. Savage complete justice, I copy the entire narrative, including his comments.

"My next story goes far beyond any of these, and — well, I will ask the reader to decide as to whether there is any help

[1] Psychics: Facts and Theories. Arena Publishing Co.

in hypnotism or clairvoyance or mind-reading, or any of the selves of the psychic, conscious or sub-conscious.

"Early on Friday morning, Jan. 18, 1884, the steamer 'City of Columbus,' *en route* from Boston to Savannah, was wrecked on the rocks off Gay Head, the southwestern point of Martha's Vineyard. Among the passengers was an elderly widow, the sister-in-law of one of my friends, and the mother of another.

"This lady, Mrs. K., and her sister, Mrs. B., had both been interested in psychic investigation, and had held sittings with a psychic whom I will call Mrs. E. Mrs. B. was in poor health, and was visited regularly for treatment on every Monday by the psychic, Mrs. E. On occasion of these professional visits, Mrs. B. and her sister, Mrs. K., would frequently have a sitting. This Mrs. E., the psychic, had been known to all the parties concerned for many years, and was held in the highest respect. She lived in a town fifteen or twenty miles from Boston. This, then, was the situation of affairs when the wreck of the steamer took place.

"The papers of Friday evening, January 18, of course contained accounts of the disaster. On Saturday, January 19, Dr. K., my friend, the son of Mrs. K., hastened down to the beach in search of the body of his mother. No trace whatever was discovered. He became satisfied that she was among the lost, but was not able to find the body. Saturday night he returned to the city. Sunday passed by. On Monday morning, the 21st, Mrs. E. came from her country home to give the customary treatment to her patient, Mrs. B. Dr. K. called on his aunt while Mrs. E. was there, and they decided to have a sitting, to see if there would come to them anything that even purported to be news from the missing mother and sister. Immediately Mrs. K. claimed to be present; and, along with many other matters, she told them three separate and distinct things which, if true, it was utterly impossible for either of them to have known.

"1. She told them that, after the steamer had sailed, she had been able to exchange her inside stateroom for an outside one. All that any of them knew was that she had been obliged to take an inside room, and that she did not want it.

"2. She told them that she played whist with some friends in the steamer saloon during the evening; and she further told them the names of the ones who had made up the table.

"3. Then came the startling and utterly unexpected statement, 'I do not want you to think of me as having been drowned. I was not drowned. When the alarm came, I was in my berth. Being frightened, I jumped up, and rushed out of the stateroom. In the passage-way I was suddenly struck a blow on my head, and instantly it was over. So do not think of me as having gone through the process of drowning.' Then she went on to speak of the friends she had found, and who were with her. This latter, of course, could not be verified. But the other things could be. It was learned, through survivors, that the matter of the stateroom and the whist, even to the partners, was precisely as had been stated. But how to verify the other statement, particularly as the body had not been discovered?

"All this was on Monday, the 21st. On Tuesday, the 22d, the doctor and a friend went again to the beach. After a prolonged search among the bodies that had been recovered, they were able to identify that of the mother. And they found the right side of the head *all crushed in by a blow*.

"The impression made on the doctor, at the sitting on Monday, was that he had been talking with his mother. The psychic, Mrs. E., is not a clairvoyant, and there were many things connected with the sitting that made the strong impression of the mother's present personality. In order to have obtained all these facts related under numbers 1, 2, and 3, the psychic would have had to be not only clairvoyant, but to have gotten into mental relations with several different people *at the same time*. The reading of several different minds at once, and also clairvoyant seeing, not only of the bruised head, but of the facts that took place on the Friday previous (this being Monday), — all these multiplex and diverse operations, going on simultaneously, make up a problem that the most ardent advocate of telepathy, as a solvent of psychic facts, would hardly regard as reasonably coming within its scope.

"Let us look at it clearly. Telepathy deals only with occurrences taking place at the time. I do not know of a case where clairvoyance is even claimed to see what were once facts, but which no longer exist. Then there must have been simultaneous communication with several minds. This, I think, is not even claimed as possible by anybody. Then let it be remembered that Mrs. E. is not conscious of possessing either telepathic or clairvoyant power. Such is the problem.

"I express no opinion of my own. I only say that the doctor, my friend, is an educated, level-headed, noble man. He felt sure that he detected undoubted tokens of his mother's presence. If such a thing is ever possible, surely this is the explanation most simple and natural."

Here, then, we have a case involving all the difficulties which stand in the way of accepting telepathy as a universal solvent for this class of mysteries. It is a representative case of the class specifically mentioned both by Mr. Savage and by Mr. Myers as being inexplicable by reference to telepathy; for the message contained information not in the possession of the psychic, or of the client, or of any of the living friends of either. Consequently, if this case can be explained by reference to telepathy, clearly science will have a right to demand further proofs of spirit communion.

"Let us look at it clearly." In order to do so we must first divest the case of several imaginary difficulties with which Mr. Savage seems to have environed it. For instance, he says that —

"In order to have obtained all these facts related under numbers 1, 2, and 3, the psychic would have had to be not only clairvoyant, but to have gotten into mental relations with several different people *at the same time*."

It is not entirely clear why it was necessary for "all these multiplex and diverse operations" to be "going on simultaneously," or, indeed, why it should be necessary for all of them to go on at all. Nor does Mr. Savage throw any very clear light upon that question. On the contrary, he seems to have propounded an irrelevant problem, which he regards as insoluble, with the view of claiming a triumph if his readers fail to solve it.

There are two groups of facts in this case which present themselves for consideration. One group consists of the facts which took place prior to the wreck; namely, the matter of the stateroom, the whist, and the partners. The other group comprises the particular circumstances attending the death of the lady.

Now we are told that it was necessary for the psychic to see clairvoyantly "not only the bruised body, but the facts that took place on the Friday previous." Mr. Savage holds the latter to be impossible for the reason that clairvoyance cannot "see what were once facts, but which no longer exist." In this proposition he is clearly right. But he would doubtless admit that clairvoyance is equal to the perception of the condition of the bruised body; and from that condition the manner of death, exactly as detailed in the message, could be clearly inferred. After all, that is the only fact which we are called upon to explain, since the facts that took place on the Friday previous were all known to surviving friends of the deceased; and the presumption is that they were all detailed to the son, who was frantically searching among the survivors for information which would throw light upon the fate of his mother. In the absence of any evidence or statement to the contrary, even by implication, this must be presumed. Clearly, then, it would not be necessary for the psychic to read "several different minds at once" in order to ascertain all the facts that took place on Friday, since those facts were in the mind of the son, and he was in presence of the psychic. But suppose that the facts had not yet been detailed to the son, it still does not involve the necessity for the psychic to be "in simultaneous communication with several minds," since telepathic communion with any one of the survivors would have put her in possession of all the facts that occurred on Friday. That supposition would certainly do less violence to the principles of scientific inquiry than it does to postulate a supermundane origin for the phenomenon.

I submit, therefore, that all the facts were easily ascertainable by the psychic by the exercise of clairvoyance and telepathy, each in its simplest and most direct mode of manifestation. The two powers, if both exist, are certainly not incompatible with each other. Indeed, they are so closely related that no one has yet been able to locate the boundary-line between them. They are divided only by their definitions.

I think that Mr. Savage will agree with me that, in the foregoing view of the case, my interpretation is to be preferred to his on the broad ground of inherent probability, since his explanation ascribes a supermundane origin to the phenomena, whilst mine ascribes it to those natural powers ot the human mind the existence of which he freely admits. Nor does it fortify his view of the case to say that the psychic does not possess either telepathic or clairvoyant power, for that is simply begging the question. The very point in controversy is whether the phenomena of so-called spiritism proceed from spirits of the dead, or are the result of the exercise of the known powers of the living. And to say, as Mr. Savage does, that the psychic "is not conscious of possessing either telepathic or clairvoyant power," is but another way of saying that this particular psychic believes that the messages transmitted through her come from disembodied spirits. It is a common thing among spiritistic psychics to disclaim telepathic or clairvoyant powers; and it is doubtless honestly done. Their theory is that their phenomena are produced by spirits of the dead, and that clairvoyance and telepathy have nothing whatever to do with it; in other words, that they "are not conscious" of possessing those powers. But when that statement is employed as an argument in support of the spiritistic hypothesis, it becomes a gross and palpable case of *petitio principii.*

Thus far I have argued this case from Mr. Savage's own standpoint; that is, I have invoked the aid of those powers of the mind, and those only, which he *knows* men to possess, namely, telepathy and clairvoyance. I have shown how these alleged powers may have operated, each in its legitimate sphere of activity, to produce the phenomena he describes. He declares that he is in search of a working hypothesis which will explain these phenomena without the necessity of invoking supermundane agencies, but has thus far failed to find one. I have herein advanced one, the fundamental postulates of which are his. They are not mine. He believes in telepathy, and I agree with him. He believes in independent clairvoyance. I do not. He says that he *knows*[1] that clairvoyance exists as a power of the human mind. I wish I possessed the same positive information. It would be a great simplifier of explanations, — a short cut across a labyrinthine field. Like the hypothesis of spiritism, it is " simpler " than the scientific explanation of the phenomena; but its chief merit consists in the fact that it saves the trouble of thinking. I have looked in vain for indubitable evidence of the reality of the power of independent clairvoyance. I do not say that it does not exist. I do not know. I simply say that I have not yet been made acquainted with facts sufficient to remove the question from the domain of doubt and uncertainty. I do know that many phenomena which a few years ago were attributed to clairvoyance are now easily explicable by reference to telepathy; and I know that the field of the former is constantly narrowing, whilst that of the latter is correspondingly widening. I know that telepathy is a faculty of the human mind; and I feel safe when dealing with that proposition. But until the boundary-line between telepathy and clairvoyance is defined with sufficient exactitude to demonstrate that there is any line at all, I shall not attempt to offer clairvoyance as a final explanation of any phenomena whatever.

Now let us examine the phenomena presented by Mr. Savage from another point of view. I will begin by quoting a proposition of his, which, if true, disposes of the case

[1] Psychical Review, vol. i. no. 1.

at once, and leaves him master of the situation. It is this, —

"Telepathy deals only with occurrences taking place at the time."

If this proposition is true, Mr. Savage and Mr. Myers are both right in rejecting telepathy as a solvent for the mystery surrounding a very large class of cases. Whether it is true or not, it clearly defines their attitude and reveals the ground upon which they stand. In saying this I do not wish to be understood as holding Mr. Myers responsible for the opinions of Mr. Savage; but as their conclusions are identical, I assume that they have reasoned from the same premises. I am confirmed in that belief for the reason that I can see no other possible ground for their conclusion.

Be that as it may, the fact remains that the assertion that "telepathy deals only with occurrences taking place at the time" is a fundamental error of the most pronounced character. It is difficult to imagine an error that could be more misleading to the searcher after truth in the psychical realm, and hence more mischievous in its consequences, than this one is.

Again, "let us look at it clearly." What is telepathy? It has already been defined as "the means of communication between subjective minds." In other words, it is the means of conveying information from one subjective mind to another. That it is confined in its operations to the subjective intelligence will not be disputed. It has already been sufficiently explained, and requires no further remark in this connection, but will be demonstrated later on. Now, when a message is telepathically sent from one subjective mind to another, it conveys some item of information to the subjective mind that receives it. That information is henceforth a part of the mental equipment of the percipient's subjective mind, and, since the memory of the subjective mind is practically perfect, it is not likely to forget any important fact that may have been thus received. But suppose the percipient does not happen, at the time of the subjective reception of the message, to be in that peculiar mental condition required to enable him to elevate his subjective impressions above the threshold of his normal, or objective, consciousness. Is the message any the less a part of his subjective mental equipment? In other words, does a failure to become objectively conscious of the reception of the message delivered to the sub-conscious intelligence cause the latter instantly to forget the subject-matter of the message? No one who is acquainted with the salient characteristics of the subjective intelligence, as developed in experimental psychology, will claim that it does. Suppose, then, that the recipient of the message does not belong to that class of sensitives who are able to elevate their subjective impressions above the threshold of consciousness. Then, suppose that a week later the recipient happens to be in the presence of a mind-reader, and they hold a séance for the purpose of making experiments in telepathy. Is there any *a priori* reason why the telepathist should not be able to read that message as it exists, latent, in the mind of the recipient? If not, why not? If he can do so, the assumption that "telepathy deals only with occurrences taking place at the time" must be revised; and with its revision the whole fabric which has been so industriously built upon that foundation must fall. The only possible resource is to deny the proposition that a message received telepathically from one source can also be delivered by the same means to a third person.

It will be necessary for us to discuss this point briefly, for it will eventually be seen that the whole fabric of spiritism, scientifically and logically considered, rests upon a tacit denial of this proposition. My proposition is this: —

If A can communicate a fact telepathically to B, it follows that B can communicate the same fact telepathically to C.

This seems to be a self-evident proposition; and no one, to my knowledge, has ventured specifically to deny its truth. Yet, as before intimated, if it is true, all of those difficulties vanish which Messrs. Savage and Myers experience in finding a telepathic explanation for their phenomena.

"I have had," says Mr. Savage, "communication while sitting in my study concerning things that were taking place two hundred miles away." And he declares that it seems to him to be "stretching the theory of telepathy and of clairvoyance beyond probability to call them in to explain" the fact. Why he thinks the explanation is outside the domain of clairvoyance, he does not tell us. It seems to me that if there is such a power as independent clairvoyance, the clairvoyant explanation is easy and perfect. But as I do not yet admit the genuineness of that power, I will try to help Mr. Savage to a telepathic explanation. He does not give us an account of the circumstances of his case; but as it is generic, I will furnish a specimen that covers the ground. The following case was related to me by a lady now living in Washington, for whose veracity and exalted character I can vouch.

Some years ago she was residing in the interior of Pennsylvania. On one occasion she visited friends in Philadelphia, and during her stay she was induced to call on a then celebrated "medium." Amongst other things the medium told her that one of the children of the lady's family had just fallen from a tree and was apparently badly hurt. The statement did not make much impression on her, for she was rather sceptical on the subject of spirit communication; but when she went home she learned that it was literally true, and the hour corresponded very nearly to the time when she was at the séance. It transpired that the child was not badly hurt, although it was insensible when picked up.

Mr. Savage would hold that it would be "stretching the

theory of telepathy" to the breaking point to call it in to explain this case. Why? Is it not probable that the lady was in telepathic *rapport* with her family at home? And would it be stretching the theory of telepathy too far to suppose that she would be informed by that means of the disaster happening to one of her family? I think not. But the lady was not a psychic, and the message, consequently, remained below the threshold of her normal consciousness. The "medium," however, was able to read her mind, and he found the fact recorded there as stated.

In a work comprising two large volumes, entitled "Phantasms of the Living," of which Mr. Myers is one of the authors, there are hundreds of cases recorded where telepathic messages were received informing the percipient of danger or disaster to loved ones at a distance. Mr. Myers and Mr. Savage will both agree that it would be in perfect accordance with the experience of mankind to suppose that the lady was telepathically informed of the accident. They will both agree that the message might have been received subjectively, and yet not brought within the domain of her normal consciousness; for that is also in strict accordance with the known facts of telepathy. Thus far we all travel along together very comfortably and harmoniously. But when the psychic imparts the information to the lady, the crucial question at once arises, "How did he obtain it?"

Here are the two hypotheses: —

Mine is that he read the mind of the lady; in other words, he obtained telepathically the information that Messrs. Savage and Myers will both admit was, or might be, legitimately in her possession through telepathic agency.

Their theory is that the spirit of some dead man was watching the child when it fell, and that he hastened to Philadelphia to hunt up a "medium" of communication between himself and the lady, so that he could tell her all about it. By a happy coincidence he found the lady and a suitable "medium" already in consultation, looking for a test case upon which to postulate a theory of immortality. And it was forthcoming. According to Messrs. Savage and Myers, it was the crucial test, — demonstrative of spirit intercourse.

If my hypothesis "stretches the theory of telepathy" too far, and if evidence of immortal life consists in the adoption of their theory of explanation, well may we exclaim, —

> "On what a slender thread
> Hang everlasting things!"

I ask the intelligent, unprejudiced reader to judge for himself which of the two explanations is more likely to be correct. To this end he must ask himself whether it is more rational to suppose that the lady obtained a telepathic message from home and transmitted the same to the psychic, than it is to suppose that it required the intervention of a supermundane agency to convey the information.

In answering this question the logical and scientific axiom must not be lost sight of, — that we have neither occasion nor logical right to seek a supermundane explanation of a phenomenon, when it is explicable by reference to natural laws with which the world is acquainted.

On this point the truly scientific reader will doubtless prefer to stand with Mr. Podmore, one of the Secretaries of the Society for Psychical Research, who says: "When the choice of explanation seems to lie between telepathy and some faculty even more dubious and more remote from ordinary analogies, it is right that the hypothesis of telepathy should be strained — if necessary, to the breaking point — before we invoke a stage-deity to cut the knot." [1]

[1] Apparitions and Thought-Transference, pp. 369, 370.

CHAPTER V.

SPIRITISTIC PHENOMENA (*continued*).

Experimental Telepathy. — Deferred Percipience. — Cases in Point. — Planchette. — Latency of Telepathic Impressions. — Nebuchadnezzar's Dream. — Daniel's Telepathic Power. — Final Explanation of Mr. Savage's Test Case. — The Mother's Message to her Son. — The Son's Message to the Psychic. — The Last Resource of Spiritism. — Mr. Savage's Crucial Question. — The Unscientific Attitude of Spiritists. — Thunder considered as the Voice of an Angry God. — The Simplicity of Nature's Laws. — The Alleged "Simplicity" of the Spiritistic Hypothesis. — It saves Thinking. — Reasoning in a Circle. — Why cannot Spirits communicate with the Living? — Not a Pertinent Question. — The Real Question is, Do they so Communicate? — The Evidence is against the Spiritistic Hypothesis. — "Spirits of Health and Goblins Damned."

I HAVE thus far examined Mr. Savage's test case from a theoretical standpoint. My theories, however, have all been based upon the well-known facts of experimental psychology, except where I have argued from a provisional assumption of the reality of the power of independent clairvoyance. I now approach the domain of ascertained facts. My text is still his declaration that "telepathy deals only with occurrences taking place at the time."[1] If it had been stated that "telepathy deals only with occurrences taking place at the time of the delivery of a message concerning them to the subjective mind of the party for whom it is intended," it would have been much nearer the truth, but would still have been far from accurate, as will be seen hereafter. Thus limited, however, it could not have been

[1] M. J. Savage in "Psychics: Facts and Theories."

pressed into the service of spiritism; and we must, therefore, presume that the words were intended and used in their full significance. In other words, the fate of the argument must depend upon the correctness of the premises as they are formulated.

In making this statement some very important facts set forth in "Phantasms of the Living," must have been forgotten for the moment, or else the article from which quotation was made was written before the publication of that voluminous record of telepathic experiences.

Be that as it may, one very important feature of the phenomena of telepathy has certainly been ignored. It is a feature, too, of the first importance, for, without including it as a factor in any given case, one is more than likely to be led into the most grievous errors. I refer to the phenomenon of "deferred percipience." The meaning of the term is thus explained by Mr. Myers in his learned and able introduction to "Phantasms of the Living": "We find in the case of phantasms corresponding to some accident or crisis which befalls a living friend, that there seems often to be *a latent period before the phantasm becomes definite or externalized* to the percipient's eye or ear. Sometimes a vague *malaise* seems first to be generated, and then when other *stimuli* are deadened, — as at night or in some period of repose, — the indefinite grief or uneasiness takes shape in the voice or figure of the friend who in fact passed through his moment of peril some hours before." He then goes on to say that "it is quite possible that a deferment of this kind may sometimes intervene between the moment of death and the phantasmal announcement thereof to a distant friend."

This is a very general, though a very accurate, statement of a principle which will presently be seen to be a corollary of the doctrine of duality of mind and of sub-conscious intelligence.

A person in imminent and deadly peril telepathically conveys a message to his nearest friend or relative, informing him of the occurrence. This may be done by means of a vision or by clairaudience, or otherwise; but it must necessarily be done by some means that addresses itself to the sensory experience of the percipient. It is a message from the subjective mind of the "agent" to that of the "percipient." If the percipient is a psychic, he will probably perceive the import of the message at once. If he is not a psychic, or is not easily thrown into the psychical or subjective condition, he may not be able for hours to elevate the message above the threshold of his own consciousness. If he is incapable (as most people are) of becoming objectively conscious of what is going on in his subjective mind, he may never be able to become normally conscious of the message that is lying "latent" in his "subliminal consciousness." Nevertheless the information is there, although he may not, as before remarked, be conscious of it at the time of its reception. It may remain latent for a week or a month; or he may never be able to take objective cognizance of it unaided by some one more sensitive to subjective impressions.

It must be remembered that telepathy is one of those psychic powers that is seldom, if ever, acquired by persons who are in a normal state of physical health. Let me not be misunderstood on this point. When we speak of one possessing telepathic power, we usually mean, simply, that he is one who is capable of taking objective cognizance, or becoming objectively conscious, of the messages received by his subjective mind. In other words, he is one who is capable of elevating the impressions of his sub-conscious intelligence above the threshold of his normal or objective consciousness. The fact that he is unable to do this is no evidence that he is incapable of receiving subjective impressions, or that he does not receive telepathic messages.

Indeed, the facts show that there is practically little difference, other things being equal, in the capacity of persons of average intelligence for receiving telepathic communications. The difference consists, not in the ability to receive, but in the ability to perceive, or to become objectively conscious, of what has been received. And the latter power usually finds its origin in an abnormal physical condition, ranging in intensity from that of an incipient neurosis to the terrible affliction endured by the Seeress of Prevorst,—the power and the physical abnormality nearly always sustaining perfectly harmonious proportional relations.

It follows that a perfectly normal, healthy man is seldom able to assimilate the full content of a telepathic message. It reaches his consciousness, if at all, only in the form of a vague impression, creating a transient feeling of unrest or foreboding, but which is soon submerged or thrown off by his superabundant vitality. Few are entirely exempt from such impressions, and they vary in intensity in proportion to their importance to the individual. But the fact that one is not able to take objective cognizance of their full import does not prove that the information, in all its details, is not indelibly stamped upon the tablets of the soul. From this postulate it follows that the work of a trained psychic, capable of reading the minds of his sitters, is all that is necessary to reveal the full content of a telepathic message latent in the subjective mind of his client.

The foregoing propositions seem almost self-evident to the merest tyro in psychic science; but as Mr. Myers and his colleagues, Messrs. Gurney and Podmore, have taken pains not only clearly to define "deferred percipience" and note it as an important factor in telepathy, but to demonstrate it experimentally and print accounts of its illustrative cases occurring spontaneously, it becomes our duty to present a few of the most prominent of those facts recorded by them, to the end that a most important factor in telepathy may not be in danger of being overlooked.

The first case to which I invite attention was experimental. The facts seem trivial; but, as they illustrate an important principle, space cannot be refused for their reproduction. The author introduces the experiment by the following prefatory remarks: —

"I will quote one more taste-series, for the sake of illustrating a special point, — namely, the *deferment* of the percipient's consciousness of the sensation until a time when the agent had himself ceased to feel it. This fact is of great interest, on account of the marked analogy to it which we shall encounter in many of the *spontaneous* telepathic cases.

"June 11th, 1885.

"Dr. Hyla Greves was in contact with Miss Relph, having tasted salad oil.

"Miss Relph said: 'I feel a cool sensation in my mouth, something like that produced by sal prunelle.'

"Mr. R. C. Johnson in contact, having tasted Worcestershire sauce in another room.

"'I taste something oily; it is very like salad oil.' Then a few minutes after contact with Mr. Johnson had ceased, 'My mouth seems getting hot after the oil.' (N. B. Nothing at all had been said about the substances tasted either by Dr. Greves or Mr. Johnson.)

"Dr. Greves in contact, having tasted bitter aloes.

"'I taste something frightfully hot . . . something like vinegar and pepper. . . . Is it Worcestershire sauce?'

"Mr. Guthrie in contact, also having tasted bitter aloes.

"'I taste something extremely bitter, but don't know what it is, and do not remember tasting it before. . . . It is a very horrid taste.'"[1]

The next experimental case is also seemingly trivial, but is important as an illustration of deferred percipience. This experiment was one of a series made by Rev. P. H. Newnham, Vicar of Maker, Devonport, England, his wife being the mind-reader. The questions were written down

[1] Phantasms of the Living, vol. i. p. 56.

by Mr. Newnham, the wife knowing nothing of their character; and the answers were written out by her by means of planchette: —

"'What name shall we give to our new dog?'

"A. 'Yesterday was not a fair trial.'

"'Why was not yesterday a fair trial?'

"A. 'Dog.'

"And again: —

"'What do I mean by chaffing C. about a lilac tree?'

"A. 'Temper and imagination.'

"'You are thinking of somebody else. Please reply to my question.'

"A. 'Lilacs.'

"Here a single image or word seems to have made its mark on the percipient's mind, without calling any originative activity into play; and we thus get the naked reproduction. In these last examples we again notice the feature of *deferred* impression. The influence only gradually became effective, the immediate answer being irrelevant to the question. We may suppose, therefore, that the first effect took place below the threshold of consciousness."[1]

To these remarks is added a footnote, as follows: —

"The following case, though not strictly experimental, is sufficiently in point to be worth quoting. Though unfortunately not recorded in writing at the time, it was described within a few days of its occurrence to Mr. Podmore, who is acquainted with all the persons concerned. The narrator is Miss Robertson, of 229 Marylebone Road, W.

"About three years ago I was speaking of planchette-writing to some of my friends, when a young lady, a daughter of the house where I was spending the evening, mentioned that she had played with planchette at school, and that it had always written for her. Thereupon I asked her to spend the evening with me, and try it again, which she agreed to do. On the morning of the day on which she had arranged to come to me, her brother, on leaving the house, said, laughing, 'Well, Edith, it is all humbug; but if planchette tells you the name and sum of money which are on a check which I have in my pocket, and which I am going to cash for mother, I will believe there is

[1] Op. cit., pp. 70, 71.

something in it.' Edith, on her arrival at my house in the evening, told me of this, and I said, 'We must not expect that; planchette never does what one wants,' or words to that effect. A couple of hours after, we tried the planchette, Edith's hand alone touching it. It almost immediately wrote, quite clearly, —

'I. SPALDING, £6 13. 4.'

I had forgotten about the check, and I said, 'What can that mean?' Upon which Edith replied, 'It is H.'s check.' I was incredulous, having a long acquaintance with planchette. I said, 'If it is right, send me word directly you get home. I am sure it will not be.' But the next day I received a letter from Edith, telling me that she had astonished her brother greatly by telling him the name and amount on the check, which was perfectly correct. I have read this account to the young lady and her brother, who sign it as well as myself.

'NORA ROBERTSON.
'E. C.
'D. C. H. C.'

"In answer to an inquiry, Miss Robertson adds, on Feb. 12, 1885:—

"'Miss E. C. says, in answer to your question, that she is quite certain she could not have known or surmised the name and amount of the check.

"'I can confirm her on the first point, for I remember questioning everybody all round at the time. She had just returned from school, and knew nothing at all about her mother's business or money matters.'

"Here, it will be observed, the impression seems not only to have been unconscious, but to have remained latent for several hours before taking effect; for it is, at any rate, the most natural supposition that the transference actually occurred at the time when the conversation on the subject took place between the brother and sister."

The intelligent reader will not fail to notice that the foregoing is not only illustrative of deferred percipience, but is representative of a very large class of cases where the message never reaches the normal consciousness of the percipient except through extraneous means. In this case it was by planchette, in the hands of the percipient; but does any one suppose that if some one else had operated the instrument, who was also a good telepathist and *en rapport* with Edith, the same result would not have been produced? For instance, suppose Miss Robertson had been thus endowed, would it have been necessary to attribute the phenomenon to supermundane agency if she had succeeded in reading in Edith's subjective mind what was obviously there, namely, the details regarding the check? Certainly not. And yet that would have been a case exactly such as Messrs. Savage and Myers have declared unaccountable except under the hypothesis of spirit communion.

The next case of deferred percipience recorded in "Phantasms of the Living" was related by the percipient, Mr. Frederick Wingfield, of France. The essential part of the narrative is the following: —

"'On the night of Thursday, the 25th of March, 1880, I retired to bed after reading till late, as is my habit. I dreamed that I was lying on my sofa reading, when, on looking up, I saw distinctly the figure of my brother, Richard Wingfield-Baker, sitting on the chair before me. I dreamed that I spoke to him, but that he simply bent his head in reply, rose, and left the room. When I awoke, I found myself standing with one foot on the ground by my bedside, and the other on the bed, trying to speak and to pronounce my brother's name. So strong was the impression as to the reality of his presence, and so vivid the whole scene as dreamt, that I left my bedroom to search for my brother in the sitting-room. I examined the chair where I had seen him seated, I returned to bed, tried to fall asleep in the hope of a repetition of the appearance; but my mind was too excited, too painfully disturbed, as I recalled what I had dreamed. I must have, however, fallen asleep towards the morning; but when I awoke, the impression of my dream was as vivid as ever, — and I may add is to this very hour equally strong and clear. My sense of impending evil was so strong that I at once made a note in my memorandum book of this "appearance," and added the words "God forbid."

"'Three days afterwards I received the news that my brother, Richard Wingfield-Baker, had died on Thursday evening, the

25th of March, 1880, at 8.30 P. M., from the effects of the terrible injuries received in a fall while hunting with the Blackmore Vale hounds.'"[1]

Following this case, and the corroborative evidence verifying it, may be found the following very sensible and judicious remarks: —

"It will be seen here that the impression followed the death by a few hours, — *a feature which will frequently recur.* [The italics are mine.] The fact, of course, slightly detracts from the evidential force of a case, as compared with the completely simultaneous coincidences; inasmuch as the odds against the *accidental* occurrence of a unique impression of some one's presence within a few hours of his death, enormous as they are, are less enormous than the odds against a similar accidental occurrence within five minutes of the death. But the deferment of the impression, though to this slight extent affecting a case as an item of telepathic *evidence, is not in itself any obstacle to the telepathic explanation.* We may recall that in *some of the experimental cases the impression was never a piece of conscious experience at all;* while in others the latency and gradual emergence of the idea was a very noticeable feature. This justifies us in presuming that an impression which ultimately takes a sensory form may *fail in the first instance to reach the threshold of attention.* It may be unable to compete, at the moment, with the vivid sensory impressions, and the crowd of ideas and images, that belong to normal seasons of waking life; and it may thus remain latent till darkness and quiet give a chance for its development. This view seems, at any rate, supported by the fact that it is usually *at night* that the delayed impression — if such it be — emerges into the percipient's consciousness. It is supported also by analogies which recognized psychology supplies. I may refer to the extraordinary exaltation of memory sometimes observed in hypnotic and hystero-epileptic 'subjects;' or even to the vivid revival, in ordinary dreaming, of impressions which have hardly affected the waking consciousness."

The next and last case for which room can be made is from a narrative related by a Mrs. Montgomery to the author of "Phantasms of the Living": —

[1] Op. cit., p. 199.

"February, 1884.

"'Nearly thirty years ago I lost a sister. The place where she died being at some distance, my husband went to the funeral without me. I went to bed early, and had a frightful dream of the funeral ceremony. I saw my brother faint away at the service, and fall into the grave. I awoke with the horror of the dream, just as my husband entered the room on his return from the funeral, which had taken place at least eight hours before. I asked him to tell me if anything unusual had happened, as I had had a terrible dream, and I related it. He said, 'Who in the world told you that? I never intended telling you.' I said, 'I only dreamt it. Just as you were coming in, I awoke.'"[1]

To this the author appends the following remarks: —

"Here the picture transferred to the percipient's dream was a precise and detailed one. It was of a sort which might at first seem more fitly to belong to a later class, where something of the nature of clairvoyance is suggested. *Nor would the eight hours' interval between the event and the dream be an objection to this view; for I have already mentioned that the deferment or latency of telepathic impressions is especially frequent in dream and 'borderland' cases*, as though the idea or image had been unable to compete with the vivid sensations which external realities force on the mind, and only got its chance of emerging into consciousness when the senses were closed to these contending influences. But seeing that at the moment of Mrs. Montgomery's dream her husband was just about to enter her room, with the shock of the burial scene probably fresh in his mind, it is at any rate conceivable that he then, and not the brother at the earlier time, transmitted the impression."

It may be here remarked that, if the author's interpretation of this occurrence is the true one, it demonstrates not only that Mr. Savage's postulate, that "telepathy deals only with occurrences taking place at the time," is conspicuously inexact, but that the person whom the occurrence most concerns is not necessarily the agent who transmits the intelligence.

[1] Op. cit., pp. 328, 329.

Even a clergyman may be prepared to relegate the second chapter of Daniel to the domain of romance; but the story is highly probable, as well as illustrative of the phenomenon of deferred percipience. Nebuchadnezzar's dream was strongly impressed upon his mind as being very important, yet he was not able to recall it to his objective mind. He demanded its recall by the Magi of all classes, and upon their failure to do so he decreed their death. Daniel and his three friends were included in the despotic decree. The four met and prayed over it, death staring them in the face. Prayer and the imminence of death put them in the best possible subjective condition, and when this was supplemented by sleep Daniel was in telepathic communication with the king, and, after the dream was at least twenty-four hours old, read it from the king's mind, which was its sole repository. The dream and the percipience were not concurrent, yet a clearer case of telepathy is nowhere stated.

I have now presented an outline of the most important factors in telepathy that seem to have been ignored by both Mr. Myers and Mr. Savage, and it remains to apply them to the case under consideration. The reader will readily recall the salient features of Mr. Savage's test case as given in full in Chapter III. The one fact necessary to account for, or, rather, the one problem necessary to solve, is, how could the details concerning the manner of the lady's death have been transmitted telepathically to the psychic who first informed the living that the victim was not drowned, but had been killed by a blow on the head?

It seems to me that there are two solutions of the problem, either one of which is adequate.

The first and most obvious one, in view of the facts pertaining to the phenomenon of "deferred percipience" which have just been detailed, is this: It will be remembered that the victim of the tragedy, "Mrs. K., and her sister, Mrs. B., had both been interested in psychic investigation, and had held sittings with the psychic," Mrs. E., the one who subsequently revealed the details of the tragedy at a sitting in presence of the son and the sister of the deceased. It will also be remembered that (again quoting the language of Mr. Savage) "this Mrs. E., the psychic, had been known to all the parties concerned for many years, and was held in the highest respect." It is also stated in the same paragraph that the psychic visited Mrs. B. regularly every Monday, and that on the occasion of these visits the three ladies "would frequently have a sitting."

Here, then, was the situation. The psychic and the deceased had not only been on the most intimate and friendly terms for many years, but they "frequently had a sitting" for "psychic investigation." Could a state of affairs be imagined that would make it more probable that the deceased would telepathically inform the psychic of the great crisis through which she was passing? Could conditions be imagined more favorable to the successful transmission of a telepathic message to a psychic? Surely not. But it appears that the psychic did not become cognizant of the facts until the sitting on Monday. It is superfluous to say that it was simply a case of "deferred percipience." The psychic did not happen to be in the subjective condition until the Monday's sitting, and consequently was not able to elevate the message above the threshold of her consciousness.

Could anything be more obviously within the recognized domain of telepathy? Yet here is a case where telepathy was dealing "with occurrences" not "taking place at the time." Here is a "communication of intelligence that neither the psychic nor the sitter possessed, nor ever did possess."

I am not yet through with the telepathic explanations of

this case, but have reserved the better one for the last. It may not be better for this particular case, but it will cover a larger number of other cases coming under the test conditions prescribed than the one just given. For the first explanation will not cover cases where the "agent" is not presumably in telepathic rapport with the psychic, — a condition obviously existent in this case. The solution I am now about to advance will not only cover this case, but will explain all that I have ever seen, heard, or read of, coming under the test formulas I have so often quoted.

The reader will have already anticipated me when I say that the most obviously natural thing imaginable was for the dying mother to send a telepathic message to her son, informing him of all the sad details of her death, possibly including the incidents of the Friday previous; although we are not bound to include the latter, for the reason that they could have been obtained from the survivors. The son was not a psychic, and consequently could not readily become conscious of the content of the subjective message. It was therefore necessarily a case of "deferred percipience," and he might never have become objectively conscious of the information which he subjectively possessed had he not consulted a psychic who could read his mind. He had that information latent in his subjective mind, and it was necessarily an open book to the psychic. If not, why not? As I have before remarked, the only possible alternative is to deny the proposition that *information obtained telepathically can be transmitted telepathically*. Again I ask, If not, why not?

Here, then, is the issue clearly defined, reduced to its lowest terms, and divested of all extraneous, irrelevant side issues. It is "the last ditch" of spiritism considered as a scientific question.

The proposition is that —

If A can communicate a fact by means of telepathy to B,

it follows that B can communicate the same fact by the same means to C.

To admit this proposition to be true is to yield the last stronghold of spiritism.

To deny it is equivalent to affirming that telepathy can be employed only to convey information received by some other means.

Is there any reason to suppose that telepathy is so restricted in its field of operations? Why should it be restricted to any two individuals in a group of three or more? As well might one say that the power of gravity is restricted to two of the heavenly bodies, and that because it operates between the sun and the earth it cannot operate between the sun and any other planet. As well might one assume that the moon does not shine upon the earth, since it is known that the moon derives its light from the sun.

The logical consequences of these two suppositions would be no more disastrous to the planetary universe than it is to the mental world to suppose that B cannot telepath to C because A can telepath to B. In the one case it leads to planetary chaos; in the other it leads directly and inevitably into the dark and dismal realm of superstition.

Knowledge of a fact obtained by means of telepathy is just as much a part of the recipient's stock of subjective information as knowledge of the same fact obtained in any other way. This being true, it is a corollary that the knowledge of that fact can be telepathically communicated to another person without reference to the method by which, or to the source from which, it was received.

The last resource of spiritism is to deny the truth of this proposition. Well may spiritists hesitate to admit its truth; for it is evident to them, as it is to every logical mind, that when once it is admitted that information telepathically received from one person can be telepathically communicated to another, it involves the admission that the same information may be transmitted by the same means to still another, and so on, *ad infinitum.* It is as obvious to them as it is to others, therefore, that, when once the admission is made, there exists a ready means of accounting for every conceivable case of alleged spirit communication.

Let us put the issue in another form: Let us suppose that A has recently died. His mother, B, is not a psychic, and knows nothing, objectively, of the death of her son, but happens soon after to consult a "medium," C, who knows, objectively, nothing of either A or B. Then suppose that C informs B that A is dead, adding particulars, which, if true, demonstrate positive knowledge; and it is subsequently found that the information is accurate.

Now, one of the two following conclusions must necessarily be the true one: —

1. The first is that A, while he was yet living, telepathically communicated the facts to his mother, B, and she, unconsciously receiving the message, telepathically communicated it to C.

2. The second is that the spirit of the dead A communicated the facts directly to C.

Admitting telepathy to be a power of the human mind, and admitting the phenomenon of "deferred percipience," the great question is, which of the foregoing explanations must science accept as the true one?

This is the "crucial" [1] question of Mr. Savage. Here is his "Rubicon" [2] which marks the boundary between the two worlds. Unlike the historical Rubicon, upon the banks of which Imperial Cæsar paused, it is both wide and deep; but they find a parallel only in the tremendous responsibilities assumed in venturing to cross.

The supposed case is identical with the "residuary phenomena" which Mr. Myers and Mr. Savage agree in attributing to spirits of the dead. I protest that I am unable to

[1] Psychics: Facts and Theories, p. 148. [2] Ibid., p. 149.

follow them. I do not know of any rule of logic or science which will warrant me in attributing to supermundane agency any phenomenon which can be explained by reference to known principles of natural law. Nor do I know of any rule which would warrant me in presuming a supermundane cause for a phenomenon when even cognate phenomena are explicable under principles of natural law.

Let us for a moment contemplate the "scientific" attitude which these gentlemen have assumed.

It is elementary to say that no hypothesis can be true unless it is capable of explaining *all* the phenomena pertaining to the subject-matter. In other words, if one single pertinent fact is inexplicable under an hypothesis, that hypothesis is necessarily wrong. Yet these gentlemen are attempting to base an hypothesis upon a "small residuum" of phenomena, whilst admitting that the "vast bulk" of cognate phenomena are explicable under well-known principles of natural law.

Now, one of the primary rules of scientific investigation is that when a series of phenomena has been attributed to supermundane agency, if one of that series can be shown to have been produced by mundane causes, all the rest of the series must be presumed to have the same origin until the contrary is demonstrated.

Under this rule, no true scientist or logician has a right, for one moment, to consider the spiritistic hypothesis as tenable after it has been demonstrated that one alleged spirit communication is the result of telepathic communion between the minds of living persons. Perhaps it would be asking too much to insist that the average advocate of spiritism should be governed by the letter of this harsh rule. But I do protest, in the name of outraged science, against all attempts to base an hypothesis upon a "small residuum" of phenomena.

Primitive man held that thunder was the voice of an

angry god. Science demonstrates that it is caused by electricity passing between two clouds, or between a cloud and the earth. Every school-boy now knows, when he sees a flash of lightning, that electricity is seeking an equilibrium, and that he may expect to hear thunder. He knows, too, when he hears a peal of thunder, that a flash of lightning has preceded it, although he may not have observed the lightning. It would cause some surprise to hear a modern scientist, trained to habits of close observation, strict analysis, and logical classification of phenomena, who, having once heard a peal of thunder and failed to observe the flash of lightning that preceded it, — I say it would cause some surprise to hear such a man calmly observe that "thunder is divided into three classes. The first is the result of electricity passing between clouds; the second is the result of electricity passing between the clouds and the earth. These comprise the 'vast bulk' of the phenomena; but there is a 'still smaller residuum' of thunder which cannot be thus accounted for, and must be held to be the voice of an angry god."

Yet this is precisely parallel to the attitude assumed by Mr. Myers in regard to alleged spirit communications. He divides them into three classes, and admits that the "vast bulk" of them can easily be explained either by reference to previous knowledge possessed by the psychic, or to that obtained by him through the medium of telepathy. But when he finds a case where the source of the telepathic message is not entirely obvious to his mind, he forgets his scientific training, and boldly crosses the Rubicon which marks the boundary between the realms of science and superstition.

Again, Mr. Savage, in effect, says that "it stretches the theory of telepathy beyond probability," to suppose it possible for a message received telepathically to be transmitted to another by the same means.

I have read somewhere of a "scientist" of a certain school, who said that he could very well understand that the apple which Newton is popularly supposed to have observed as it fell to the ground, may have been influenced by gravity when it performed that historic feat. He was also ready to admit that the earth — directly as to the mass and inversely as to the square of the distance — moved to meet the apple, as that would not require the earth to move very far, and consequently it was not very much of a concession, anyway; but he thought it was stretching the theory of gravitation a little too far to suppose it to be capable of reaching out into the space and influencing the sun, moon, and stars. He was one of that numerous class who are constantly descanting upon the "simplicity" of Nature's laws; and yet he preferred to believe that each planet was held in its course by a miracle.

He has many followers to-day who hold that it is "much easier, and therefore more in accordance with the simplicity of the operations of Nature," to attribute spiritistic phenomena directly to spirits of the dead than to attempt to account for them by the "complicated" theory of telepathy.

True, it is "easier" and "simpler." It saves thinking. It was also easier and simpler to suppose that the earth was flat. It complicated matters very decidedly when it was discovered that it is round; and still more when it was found that it is not the centre of the universe, but only an infinitesimal part of the stupendous whole.

There is a constant tendency in the popular mind to confound simplicity of formula with simplicity of operation. The former is generally simple to the last degree. The latter is infinitely complicated. Thus, nothing could be simpler than the formulas expressing the three laws of Kepler. But what a vast and complicated system they represent! A single instance will illustrate my meaning.

Nothing could exceed in simplicity his statement that

"the planets move in ellipses, having the sun in one focus." The old astronomers, ever in search of simplicity, conceived the notion that the planets must move in perfect circles around the sun. But Tycho Brahe's accurate observations, seconded by Kepler's genius for generalization, developed the fact that the planets move in curves of the extremest possible complexity. This contrast between the simplicity of a formula and the complexity of its application is observable in all of Nature's operations. It is true in mechanics, as shown, for instance, in the formula defining the power of the lever. That formula is so simple that a child can grasp its fundamental significance; yet in every complicated machine its powers are developed and utilized, with its various modifications, in a thousand different places and directions; and in all animated Nature it is utilized in such an infinity of complexities as to defy analysis. This contrast is as true in the realm of psychology as it is elsewhere. The formulas expressive of the greatest truths are always simple. They can be comprehended in their fundamental significance by the most ordinary intelligence; but a lifetime of study will fail to discover the infinite complications involved in their practical operations.

Simplicity of operation, therefore, is not a test of scientific truth. That notion belongs to the primitive ages of scientific investigation. On the contrary, every new discovery of a natural law reveals an infinity of unexpected complications in the operations of the forces of Nature. Nor is simplicity of statement a sure criterion of truth, although its opposite always renders a proposition open to suspicion. It is obvious that an error can be formulated in as simple terms as a truth.

Thus, when we are told that a psychic phenomenon is produced by disembodied spirits, and are gravely informed that this explanation is the simplest, and therefore more in accordance with the simplicity of Nature's laws than is a "complicated" telepathic explanation involving the exercise of that power by at least three individuals, the statement does not carry conviction to the scientific mind, "simple" as the explanation may be. The spiritistic explanation does not explain. It merely gets rid of the question by thrusting it outside the domain of science, outside the region of ascertained facts and the known laws of the human mind.

This attitude is all the more remarkable in a scientist when we consider the fact that in abandoning the realm of demonstrated laws he plunges into a region of which nothing is definitely known, — into an hypothetical world, peopled by hypothetical spirits. And it is still more remarkable when we reflect that the only fact upon which he could possibly base a claim to a logical right to enter the unknown world for an explanation, is the very fact in dispute. He does not even claim to have any fact, or class of facts, upon which to base his hypothesis of spirit communion other than the very ones in controversy. His attitude, therefore, is this: —

A phenomenon is to be accounted for. On the one hand, it is claimed that it is explicable by the well-known facts of telepathy. On the other hand, Messrs. Savage and Myers say, "No, we do not know how to explain it by telepathy. It must, therefore, be spirits of the dead." If asked what facts they have in support of the spiritistic theory, their reply is, "The fact that it cannot be explained by telepathy." How do you know that it cannot be explained by telepathy? "Because the phenomenon was produced by spirits."

In other words, such scientists reason in a circle; and their hypothesis is unsupported by anything save their bare, bald assertion that a "small residuum of phenomena" proceeds from disembodied spirits.

It may be replied that my theory rests upon the mere assertion that the phenomena are explicable by reference to telepathy. This, however, cannot reasonably be said; for my theory is in direct line with the well-known and admitted facts of telepathy, whilst theirs implies an utter abandonment of the domain of demonstrable facts. My theory is wholly within the logical limits of scientific induction, whilst theirs implies a bold leap into the realm of superstition.

The question is often asked, "If two embodied spirits can communicate with each other by means of telepathy, why cannot a disembodied spirit communicate with one still in the flesh by the same means?" My answer is, I do not know. Nor do I know of any one who does know. I submit, however, that it is not a pertinent question; for, be the facts as they may, it is obvious that no one can tell *why* disembodied spirits can or cannot communicate with the living. The real and only pertinent question is, "*Do* disembodied spirits communicate with the living?" The answer to this question must be made by each individual for himself, and the character of the answer will depend upon the evidence before him and his capacity for estimating its value. My own answer has already been given in the preceding pages, and more fully elsewhere.[1] That answer is, that there is no valid scientific evidence whatever that spirits of the dead have ever communicated in any manner with living persons. In investigating this question I have been influenced solely by a desire to learn the truth; and in coming to this conclusion I have been guided by the rules and axioms of logical, scientific investigation, as I understand them. Amongst the latter I have found none more worthy of confidence than those set forth by Mr. Myers in his admirable Introduction to "Phantasms of the Living." One sentence of that essay should be stamped upon the memory of every investigator of psychic science. It is this: "*We must not rashly multiply the problems involved in this difficult inquiry.*" This declaration refers to the very question now

[1] See "The Law of Psychic Phenomena."

under consideration; namely, whether any part of the phenomena of supersensory transference of thoughts or messages are produced by spirits of the dead. Continuing, Mr. Myers says: "It is certainly safer to inquire how far they can be explained by the influences or impressions, which, *as we know by actual experiment, living persons can under certain circumstances exert or effect on one another*, in those obscure supersensory modes which we have provisionally massed together under the title of Telepathy." This is Mr. Myers the scientist. Mr. Myers the spiritist has, nevertheless, "rashly multiplied the problems involved in this difficult inquiry" by ascribing a part of the phenomena to disembodied spirits. I have therefore appealed from Mr. Myers the spiritist to Mr. Myers the scientist,—with what success I leave our readers to judge.

I have stated all I deem it necessary to say in regard to spiritism, considered as an alleged means of communicating with disembodied spirits. I have confined my remarks to the "residuary phenomena" which embrace all that remains to be accounted for, according to the deliberate admissions of two of the ablest scientific advocates of spiritism now living. I do not expect other spiritists to be bound by their admissions; for other spiritists are satisfied with a far inferior grade and weight of evidence than they are. Indeed, I know of no one in the ranks of spiritism who is so careful as they are in weighing the value of evidence for or against the spiritistic hypothesis. Nor do I know of any whose qualitative and quantitative analysis of spiritistic phenomena has left a smaller residuum of facts upon which to base the hypothesis of spirit communication. The scientific world will never cease to be grateful to them for the painstaking care which they have exercised in eliminating the "vast bulk" of the phenomena which have been attributed to supermundane agency; and if I have succeeded in reducing them to a "still smaller residuum," I shall beg the privilege of quietly basking in the reflected glory of their achievements.

As I remarked at the opening of this discussion, I have felt compelled to treat the subject of spiritism at some length, for the reason that it is always of the first importance that the basic facts under consideration, in any scientific investigation, should be properly classified. The gentlemen whose views I have criticised will be the first to indorse this proposition. Moreover, each of us is in pursuit of the same ultimate object; namely, a scientific demonstration of a future life. The broad line of difference in our methods of reasoning up to that conclusion may be summed up briefly as follows: —

My proposition is that psychic phenomena, properly interpreted, including that which they attribute to disembodied spirits, furnish indubitable evidence of a future life; and that the only interpretation which science can give to such phenomena is that it emanates from the living psychic, and *never* from disembodied spirits.

They hold that psychic phenomena, of the so-called spiritistic variety, are valuable as evidence of a future life only on the supposition that they, or *some* of them at least, emanate directly from disembodied spirits; and that a demonstration that disembodied spirits can communicate with the living constitutes a demonstration that there is a future life for mankind.

I hold that so-called spiritistic phenomena are valuable as evidence of a future life only on the supposition that *none* of them emanate from disembodied spirits. My reasons are briefly these: —

In the first place, two antagonistic hypotheses cannot both be correct; nor can each be partly true and partly false: for any hypothesis that does not explain *all* the facts is necessarily wrong, and therefore utterly valueless. Thus, if any one of a series of so-called spiritistic phenomena can be demonstrated to emanate from disembodied spirits, the telepathic hypothesis is necessarily invalid as a solvent for that series of phenomena. On the other hand, if one of said series can be demonstrated to be referable to telepathy between living persons, the spiritistic hypothesis is necessarily wrong. In other words, it is a logical necessity that, as between two antagonistic hypotheses, one or the other must be wholly right and the other wholly wrong, or both must be wholly wrong. The nature of the case does not admit of compromise; for principles of natural law are not established by majorities of facts, nor are there exceptions in the operation of natural laws. It follows that if one psychic phenomenon could be scientifically demonstrated to have been produced by disembodied spirits, the whole subject would be relegated to logical chaos, and some solution of the mystery would have to be sought for other than that embraced in either of the hypotheses under consideration.

Moreover, if all the phenomena which have been ascribed to supermundane agency could be demonstrated to proceed from disembodied spirits, the problem of a future life would be not a whit nearer to a solution than it was when Job propounded his momentous question; for the question of spirit identity would still arise to plague the faithful. It will not be denied that the question of spirit identity is, and ever has been, the one great problem which defies solution. Nor will it be denied that, if it is true that spirits do communicate with the living, there is indubitable evidence that there are evil spirits as well as good; that there are ignorant spirits as well as enlightened; that there are "spirits of health" as well as "goblins damned;" and that their intents are sometimes "wicked" and sometimes "charitable." If, therefore, we are forced to accept alleged spirit communications as genuine emanations from disembodied spirits, it by no manner of means follows that one of them comes from a spirit who has once been incarnated; and the problem of a future life for man is just as far from a solution as it was before kitchen furniture began to testify and hysterical women to teach the science of the soul.

CHAPTER VI.

ANCIENT PSYCHIC PHENOMENA.

The Importance of Correct Classification of Phenomena. — The Science of the Soul. — The Phenomena of the Soul. — Old Testament Records. — The Pentateuch. — The Higher Criticism. — The Psychic History of the Children of Israel. — Unreasoning Scepticism. — Aaron's Rod. — Moses as a Psychic. — His Methods and his Instrumentalities. — The God of Moses. — His Human Characteristics. — His Advice to "spoil the Egyptians." — Moses' Interview with God on Mount Sinai. — The Molten Calf. — The Anger of God. — His Determination to destroy the Children of Israel. — Moses argues the Question. — He causes God to Repent. — Renewal of the Covenant. — Objective Moses *vs.* Subjective Moses.

IF my remarks thus far made have led the reader to infer that I regard the phenomena of spiritism as an unmixed evil, I hasten to remove the impression. Of no phenomenon of Nature can this properly be said. The phenomena of Nature are the facts of Nature; and it is from those facts that we must study the sciences. It is only when we wrongly interpret or erroneously classify a fact, that it bears upon its face the appearance of evil, or is divested of its importance to mankind. Every fact in Nature is important if properly classified and interpreted. Conversely, if the most apparently insignificant fact is improperly classified, it often becomes a stumbling-block of great magnitude in the pathway of the searcher after truth. No true scientist can, or will attempt to, deny the truth of these elementary propositions. My remarks relating to psychic phenomena have thus far been made for the sole purpose of properly classifying a very important series of psychic facts, — the facts of so-called spiritism. No one will deny the importance of this first step, if we are to study the subject scientifically. That we must study it scientifically if we would arrive at the truth, all will admit.

Psychology is, or should be, an exact science. There is no more reason why it should not be an exact science than there is why astronomy or electricity should be left out of that category; provided always that we apply the same rules of investigation to psychology that we apply to any of the other sciences. To that end, the one fundamental prerequisite is that we carefully and conscientiously study the facts. In doing this, however, the first thing needful is to divest ourselves of all prejudice arising from preconceived opinions of our own. The second, which is equally important, is to divest our minds of all prejudice in favor of the opinions of others whose conclusions are not based upon well-authenticated phenomena. The third, and perhaps the most important of all, is that we should divest ourselves of all emotional opinions. It is to the misdirected emotions of mankind that the world has been indebted for all the opposition that has ever been directed against the progress of science. It was the misdirected and perverted emotion of religious worship that directed the tortures and lighted the fires of the Inquisition, and has again and again drenched the earth with human blood; and no one will deny that the ranks of spiritism are to-day largely recruited from the class of people who allow their emotions to blind their judgment. It is, therefore, of the utmost importance that we should guard ourselves against the tendency to emotionalism when studying the science of psychology; for those very emotions constitute a most important part of the facts which we shall be called upon calmly to investigate.

The term "psychology" is derived from the two Greek

words *psyche*, the soul, and *logos*, a treatise. Psychology is, therefore, the science of the soul.

Like every other science, it must be studied by and through the observation of the facts pertaining to the subject treated. As the science of astronomy must be studied by observing the movements of the heavenly bodies, the science of chemistry by the study of chemical combinations and reactions, the science of geology by a study of the physical structure of the earth; so must the science of psychology be studied by and through the observation of the phenomena of the soul.

To reduce my propositions to a more orderly form, they stand thus: —

1. The Psyche, or what I have elsewhere termed the "subjective mind," is the soul.

2. The phenomena of the soul are, therefore, what are generally termed "psychic phenomena."

3. The soul is the source of all psychic phenomena.

4. The emotions of religious worship, and the longings for immortal life, are psychic phenomena.

5. It follows that the facts which we must study, and from which we must deduce all legitimate, logical conclusions relating to the science of the soul, consist of observable psychic phenomena.

It is only by an appeal to these facts that we can scientifically demonstrate that man has a soul. It is by reference to psychic phenomena alone that the existence of a Deity can be demonstrated. Physical science can do neither the one nor the other; nor can it throw more than the faintest glimmer of light upon either question.

To the materialistic scientist *physical nature conceals all that man would know of God or of his own soul. Psychic phenomena alone reveals that knowledge.*

The science of the soul is, therefore, necessarily the science of religion. To attempt to divorce religion from psychic phenomena is to attempt the impossible. It is only when psychic phenomena are misinterpreted that the cause of religion falls into disrepute, or that a knowledge of the living God is withheld from mankind.

Ignorance of psychic laws has placed gods upon the throne of heaven possessing all the frailties, weaknesses, and passions common to mankind. It has peopled the earth with "spirits of health" and with "goblins damned." It has been the source of every superstition that has ever terrified the soul and warped the judgment of man. It has created a material hell and filled it with demons, a material heaven and peopled it with demigods.

On the other hand, grossly as psychic phenomena have been misunderstood, the fact still remains that they have constituted the foundation of every religion worthy of the name. Such phenomena have been the only means whereby man has been led to the conception of a higher power. They have constituted the basis of his hope for a life beyond the grave; and they have furnished the evidence of the divine mission of all the epoch-making characters who have been instrumental in lifting the souls of men to a higher moral and spiritual plane. In a word, they have constituted the great bulwark which has protected mankind against the assaults of materialism and its consequent moral irresponsibility.

The spiritual history of man is, therefore, but a record of psychic phenomena.

In the world's intellectual infancy all the phenomena of Nature were of necessity grossly misinterpreted. Science, however, has revealed much of truth concerning the laws of the material universe. It has removed physical Nature from the domain of the supernatural, and safely conducted it within the province of induction. The future of physical science is, therefore, assured. None of its phenomena will ever again be relegated to the realms of the supernatural.

Its future is eternal progress. Spiritual science has yet to be formulated and brought within the realm of induction.

The Old Testament records furnish the most striking illustrations of what I have said of misinterpreted psychic phenomena. Much valuable time has heretofore been employed in the discussion of questions pertaining to the authenticity and the historical accuracy of the Pentateuch. Countless treatises have been written, displaying profound learning and patient research, in which are discussed questions such as whether Moses is the author of the whole, or of only a part, or, indeed, of any of the Pentateuch; whether it is an historical work, or merely legendary and poetical (De Wette); whether Ezra was its sole author, as also of all the other historical books of the Old Testament (Spinoza); or whether it is an agglomeration of fragments written by many men at different epochs (Geddes). Much ingenuity has also been displayed in attempts to prove that the whole has a mystical meaning; in which efforts its historical possibilities are entirely ignored.

The rabbins held that all but a small portion of the latter part of Deuteronomy was written by Moses. From the Jewish synagogues this belief was inherited by the Christian Church, and it is still widely prevalent among Christians of the present day. It was not until the seventeenth century that the belief was seriously challenged or the doctrine of plenary inspiration questioned by any part of the Church. The higher criticism of more recent times, however, has done much to dispel the palpable errors of former beliefs, and to bring comparative order out of legendary chaos.

It is not my purpose to enter the prolific field of discussion which these questions present. I wish, however, briefly to touch upon what seems to me to be the salient feature of the Old Testament records; namely, their character as depositories of psychic facts.

It is fashionable to deride the history of the children of

Israel and of their exodus from the land of their bondage and their sorrows. The unreasoning sceptic regards it as a fiction, written in the infancy of the human intellect, by men whose minds were dominated by superstitions, or by the priesthood for their own aggrandizement and the subjugation of their followers. Even those who believe in the plenary inspiration of the Bible and the literal truth of the Mosaic account of creation, find it hard to identify the God of love, mercy, and benevolence which Christ taught mankind to adore, with the God of Moses, who was characterized by all the weakness, passion, jealousy, cruelty, and vindictiveness common to primitive humanity. It seems evident that between the extremes of unreasoning scepticism on the one hand and of equally unreasoning credulity on the other, the truth must be found. When it is sought in that direction, it will be discovered that the questions of authorship and dates possess very little importance compared with the insight which will be gained of the first great step in the evolution of spiritual man.

As before intimated, it is from the psychical standpoint that we must study the early history of the religion of the Jews; and as it is from that religion that the religion of Christendom has been evolved, the subject possesses the most transcendent interest and importance.

If any one who understands the elementary principles involved in the production of the psychic phenomena known to this generation as "spiritistic," will, in the light of that knowledge, read the Old Testament, particularly that part of the Pentateuch relating to the exodus of the children of Israel from Egypt, he will find a record of psychic phenomena unsurpassed by any of the alleged performances of the adepts of the Orient. Viewed in that light, it will be found that many of the statements there recorded, which have provoked the scepticism and excited the ridicule of a great portion of the human family, will assume such an air of probability that they may be regarded as within the range of possible history. In saying this, I do not take into account the so-called miracle of the rod of Aaron, which, thrown upon the ground, changed into a serpent; although it is no more wonderful than many of the alleged facts of the Yogis of India which seem to be well authenticated. Indeed, those feats of Moses and Aaron were paralleled at the time by those of Pharaoh's magicians and sorcerers, whose rods also exhibited the same phenomena when thrown upon the ground; albeit the rod of Aaron demonstrated his greater power as a magician. The point is that it was the same power in the one case as in the other. If the magicians did their work within the domain of natural law, we have no right to suppose that Aaron transcended that limit. Assuming the statements to be true, the only legitimate conclusion is that Aaron was the greater magician. In other words, the same principle that I have so often tried to enforce applies here; namely, that we have no logical or scientific right to attribute any phenomenon to supernatural agency when cognate phenomena are explicable by reference to natural causes.

But it is not of the phenomena recorded in the Pentateuch where the modern parallels are involved in scientific doubt that I wish to speak. The salient feature of the psychic career of Moses consists in his supposed communion with God. Of those phenomena we have innumerable parallels in modern times which are scientifically authenticated, and we know something of the laws which pertain to their production.

That Moses was a psychic, is evident. He was familiar with all the occult arts known to the magicians and conjurers of Egypt. He had been educated partly in the household of the Egyptian royal family; but his early training was in his mother's home and family, and he was consequently imbued with a belief in the God of Israel and reverence for his name. His conception of the attributes of the Deity was necessarily limited by the prevalent beliefs of his people and the traditions of his ancestry. He was wise, energetic, and ambitious. Educated and reared in the household of Pharaoh's daughter, he was advanced far beyond his people in knowledge and practical education. He had not shared the miseries of their bondage, nor had his spirit been broken by the tyranny of their taskmasters; but his sympathies were with them, and his proud, imperious soul revolted against the oppression and degradation of his people. His mind was the storehouse of the traditions of his race, and he had faith in the promise of the God of his fathers. That promise had been a tradition handed down to the Israelites from a remote ancestry. It sustained them during all the long years of their bondage, it encouraged them during all the long and weary journey in the wilderness, and it is still a living element of the religion of that race. That Moses was thoroughly imbued with faith in that promise, is shown by the facts of his subsequently reminding God of it at critical periods in the history of the exodus, —

> "Remember Abraham, Isaac, and Israel, thy servants, to whom thou swarest by thine own self, and saidst unto them, I will multiply your seed as the stars of heaven, and all this land that I have spoken of will I give unto your seed, and they shall inherit it forever."

Moreover, Moses had committed a crime which compelled him to flee from his native land to escape the vengeance of the king. During the long years of his exile he had cherished a personal hatred of the reigning family of Egypt, for that "Pharaoh sought to slay him." Undoubtedly that feeling extended to the whole Egyptian race, and added a potent element to the combination of circumstances all of which conspired to point to him as the future deliverer of his people from their slavery. That such a chain of circumstances would naturally be so construed by such a man as the biographers of Moses have portrayed, there can be no doubt. That this was the actual effect upon his mind, cannot be questioned in view of subsequent events. His mind was filled to saturation with the auto-suggestion which crystallized in the vision which he saw on Mount Horeb, and which found voice in the command there given to go into Egypt, deliver his people from their bondage, and conduct them to the promised land. In view of the developments of modern science, there can be no other rational interpretation of the phenomenon alleged to have occurred on Mount Horeb. He was a psychic. He subjectively saw the vision of the burning bush, and he subjectively (clair-audiently) heard the voice. His education and habit of thought produced an auto-suggestion that it was the voice of God; and, true to the universal law of suggestion, it assumed to be the voice of God. The effect of this vision and of this command upon the mind of such a man as Moses could have been no other than what it is represented to have been. It gave direction, tone, and color to his whole subsequent career; and the events which followed constitute a history of psychic phenomena which find many parallels in the subsequent developments of psychic research. Doubtless many of them have been exaggerated by his historians, and many have been evolved from their inner consciousness. But enough remains, after all due allowances are made, to constitute a history the credible events of which are of the utmost value to mankind; for without them the history of the evolution of the spiritual man would not be complete.

It is not my purpose to trace in all its details the history of the psychic manifestations recorded in the Old Testament. My primary object is to suggest a line of study which cannot fail to interest the student of psychology, and which may lead to important conclusions. It is important, however, that the salient features of the communications which it is alleged that God made to Moses and the prophets should be noted; for it is by the character of the communications themselves that they must be judged. If they actually emanated from the Deity, surely there will not be lacking internal evidence to sustain that hypothesis. On the other hand, if they emanated from a finite intelligence, their character and content cannot fail to demonstrate the fact. In other words, we must apply the same standards of comparison to the psychic manifestations of Moses that we employ when estimating the value and determining the source of the messages delivered through the psychics of to-day. The most ardent spiritist, possessed of sufficient intelligence to seek shelter during the progress of a storm, would not ask us to credit the statement of a psychic that a message emanates from the spirit of a Webster, when it is couched in the language of a stevedore. Nor can we be expected to believe that a message emanates from the Deity when we find that it exhibits all the passions and frailties of our common humanity. It will not be denied by the most ardent advocate of the dogma of plenary inspiration that the God of Moses, as represented in the Pentateuch, exhibited many of the frailties and some of the vices of human nature. Nor will it be denied that this fact has been a potent weapon in the hands of scepticism in all ages of the civilized world. The contrast between the God of Moses and the God whom Jesus proclaimed to mankind is too violent to permit even the most ardent Christian to recognize their identity. Hence the resort to the theory of the symbolism of the Pentateuch on the one hand and of its mysticism on the other. It was a wise remark of Dante that, before we attempt to explain the mystical meaning of a passage of Scripture, it is well to be certain that the passage is mystic. This rule embraces a world of practical wisdom; but it is not generally followed. Hence there has been a great deal of mystical meaning extracted from the Bible where no possible mysticism was intended; and much symbolism has been drawn out of passages that are not symbolical. It may be difficult at times to determine whether or not a mystical meaning is to be drawn from a particular passage; but there is one characteristic the presence of which renders it reasonably certain that a passage is neither mystical nor symbolical. When the author assumes to be relating historical facts, and the circumstances which he details accord with the experience of mankind, it may be safely assumed that the passage is not mystical. It may be fiction, but it is neither mysticism nor symbolism. The application of this test will remove a very important part of the Old Testament outside the range of the symbolical theory. This remark applies particularly to the psychic phenomena of which we have been speaking. As these phenomena consist largely of the alleged communion of God with Moses and the prophets, I propose briefly to examine one or two examples which seem to combine the essential features of them all.

One of the most striking examples of the intensely finite, human character of the God of Moses is found in the advice given to the latter during their first interview on Mount Horeb. After commanding Moses to go to Pharaoh and demand the release of the children of Israel, and promising to conduct the latter to a "land flowing with milk and honey," he said, —

> "And I will give this people favor in the sight of the Egyptians: and it shall come to pass, that, when ye go, ye shall not go empty:
>
> "But every woman shall borrow of her neighbor, and of her that sojourneth in her house, jewels of silver, and jewels of gold,

and raiment: and ye shall put them upon your sons, and upon your daughters; and ye shall spoil the Egyptians."[1]

Comment upon the intensely human character of this advice would seem superfluous.

Perhaps the best illustration of the human characteristics of the God of Moses is found in the thirty-second chapter of Exodus. It will be remembered that when the children of Israel arrived at the foot of Mount Sinai and pitched their encampment, Moses went up into the mount to commune with God, and remained there forty days. It was then that the "tables of stone" were prepared, and "written with the finger of God."

The account of the events which happened during the long absence of Moses continues as follows: —

"And when the people saw that Moses delayed to come down out of the mount, the people gathered themselves together unto Aaron, and said unto him, Up, make us gods which shall go before us: for as for this Moses, the man that brought us up out of the land of Egypt, we wot not what is become of him.

"And Aaron said unto them, Break off the golden ear-rings, which are in the ears of your wives, of your sons, and of your daughters, and bring them unto me.

"And all the people brake off the golden ear-rings which were in their ears and brought them unto Aaron.

"And he received them at their hand, and fashioned it with a graving tool, after he had made it a molten calf: and they said, These be thy gods, O Israel, which brought thee up out of the land of Egypt.

"And when Aaron saw it, he built an altar before it; and Aaron made proclamation, and said, To-morrow is a feast to the Lord.

"And they rose up early on the morrow, and offered burnt-offerings, and brought peace-offerings: and the people sat down to eat and to drink, and rose up to play.

"And the Lord said unto Moses, Go, get thee down: for thy people, which thou broughtest out of the land of Egypt, have corrupted themselves:

[1] Exodus iii. 21, 22.

"They have turned aside quickly out of the way which I commanded them: they have made them a molten calf, and have worshipped it, and have sacrificed thereunto, and said, These be thy gods, O Israel, which have brought thee up out of the land of Egypt.

"And the Lord said unto Moses, I have seen this people, and, behold, it is a stiff-necked people:

"Now therefore let me alone, that my wrath may wax hot against them, and that I may consume them: and I will make of thee a great nation.

"And Moses besought the Lord his God, and said, Lord, why doth thy wrath wax hot against thy people, which thou hast brought forth out of the land of Egypt, with great power, and with a mighty hand?

"Wherefore should the Egyptians speak and say, For mischief did he bring them out, to slay them in the mountains, and to consume them from the face of the earth? Turn from thy fierce wrath, and repent of this evil against thy people.

"Remember Abraham, Isaac, and Israel, thy servants, to whom thou swarest by thine own self, and saidst unto them, I will multiply your seed as the stars of heaven, and all this land that I have spoken of will I give unto your seed, and they shall inherit it forever.

"And the Lord repented of the evil which he thought to do unto his people."[1]

Of course, there will always be room for a variety of opinions concerning the true interpretation of these passages of Scripture. By the unreasoning sceptic they will always be regarded as purely fictitious; although we have no logical right to dismiss them thus so long as there remains any other rational interpretation. A book which, in all ages of the civilized world, has been regarded by a large and intelligent part of the human family as sacred history, cannot be dismissed without a respectful hearing.

By many these passages will always be regarded as symbolical; although it is difficult to imagine what they can possibly prefigure, what practical lesson they can teach, what principle they can symbolize.

[1] Exodus xxxii. 1–14.

By a few they will always be regarded as literal truth, written by divinely inspired men; although it is difficult to imagine how they can reconcile the crude and primitive conceptions of God which these passages develop with that grand and noble conception of the Deity which Christ taught to mankind.

By most people, however, the Pentateuch will be regarded as a collection of traditions handed down through many generations, corrupted in their transmission, recorded by different individuals, and collected and arranged by some one or more writers not definitely identified. That this is the true hypothesis, there can be little doubt, in view of the developments of the higher criticism of modern times.

There is, however, always truth in tradition. No matter how grossly the original story may have been corrupted in its transmission from mouth to mouth through the ages of its life, the salient feature of a national tradition always retains its identity and essential character. The imagination of those through whom it passes may add details embodying their own notions of what ought to have been true, and the story may thus grow in magnitude, lose coherency, and possibly become a grotesque caricature of the original. But the central idea retains its identity, and, to a certain extent, its consistency; for its foundation is generally some important truth. And the vitality of truth is such that it can never be wholly extinguished, however thickly it may be overlaid with error. "The eternal years of God are hers."

Now the central truth of the Mosaic traditions consists of the fact that Moses believed himself to be in direct communion with God. The fact that he was in error regarding the source of his communications does not militate against the verity of the tradition. He believed it to be God, and his followers so believed; and to them it was a most vital truth, for it gave tone and color and substance to the Jewish nation. It was during those communications that the covenant was from time to time renewed, — the covenant which God had made with Abraham and with Isaac and with Jacob, that their seed should "multiply as the stars of heaven," that they should be his "chosen people," and that the "land of Canaan should be their inheritance forever." It was their faith in this covenant that sustained them in every adversity and filled them with a pride and a hope which has not yet ceased to be a vitalizing element in their national and religious character. It was inevitable, therefore, that the central idea of their national tradition should be preserved. It was also more than likely that the main feature of the methods and of the paraphernalia employed by Moses in carrying on his intercourse with God should be preserved practically intact in the national tradition. All the other facts and alleged facts of Jewish history up to that time were utterly insignificant beside the one central idea that God had appeared unto Moses, talked habitually and familiarly with him, and had covenanted with them to give them the land of Canaan for an inheritance forever, and to make of them a great nation. It is a corollary of this postulate that the phenomena which Moses and his followers attributed to divine agency actually occurred, substantially as they are related in the Old Testament.

This view of the case will be still further confirmed when we consider the specific character of the phenomena and compare it with cognate phenomena which are occurring every day and which can be experimentally reproduced. In making this comparison it must be remembered that Moses was subject to the same laws, and was hedged about by the same limitations, that control and limit the psychic manifestations of to-day. The same law of suggestion operated to cause his subjective mind to believe itself to be God, that causes that of the modern psychic to believe itself to be the spirit of any deceased person whose name is suggested. If there could be any possible doubt of the truth of this proposition, it will be set at rest when we consider the nature of the conversation detailed in the foregoing passages. God is there represented as being so deeply moved by anger and jealousy, when he learned that the children of Israel were worshipping other gods, that he wanted to wipe them all out of existence. "Let me alone," said he to Moses, "that my wrath may wax hot against them, and that I may consume them." I submit that this is not the language of a God. But when we remember that the subjective mind is the seat of the emotions, and that it is egotistical, vain, selfish, and jealous to the last degree when uncontrolled by objective reason, we have no difficulty in tracing the expression to the subjective mind of Moses himself. This view is still further confirmed by the attitude of objective Moses. His reason told him that it would be highly injudicious, to say the least, to utterly consume the children of Israel; and accordingly he proceeded to argue the question with God and advise him against such heroic measures. So cogent were his arguments, or, to speak in modern scientific phrase, *so potent were his suggestions*, that God is represented to have "repented of the evil which he thought to do unto his people."

It needs no argument to convince the intelligent reader of the absurdity involved in the supposition that finite Moses was able by argument to convict of wrong-doing a God of infinite intelligence, mercy, wisdom, goodness, and power, and cause him to repent of his evil intentions. It is self-evident that the only rational explanation is that given by Maimonides, who flourished in the twelfth century, and was doubtless the greatest Jewish philosopher the world has ever seen; namely, "It was objective Moses talking with subjective Moses."

This is certainly the only explanation that will harmonize all the alleged facts and give coherency and consistency to the Old Testament accounts of the intercourse of God with man during the Mosaic dispensation. The same hypothesis applies with equal force to the intercourse of God with the prophets and seers, from the days of Abraham to the advent of Jesus of Nazareth.

Studied from this point of view, the facts related will be found to be illustrative of the principles and laws which modern scientific research has brought to light. As I have before remarked, the Old Testament is a record of most remarkable psychic experiences, — a vast storehouse of misinterpreted and wrongly classified psychic facts. But, as I shall attempt to show in the ensuing chapters, they are facts which, properly classified and intelligently appreciated, are of the most transcendent interest and importance to mankind.

CHAPTER VII.

ANCIENT PSYCHIC PHENOMENA (*continued*).

The Prophets of Israel. — Elisha's Methods. — He saves the Three Kings. — Human Characteristics of Elisha's God. — The Evolution of the Monotheistic Idea through Psychic Phenomena. — The First Conception of the Idea of a Living God. — The Evolution of the Spiritual Man. — The First Great Step through Psychic Phenomena. — The Jewish Origin of Monotheism — The God of Abraham. — The Dispensation of Moses. — The Second Great Step in the Evolution of the Spiritual Man. — The Decalogue. — The Influence of Egyptian Civilization. — The Wisdom of Moses. — Egyptian Ethics and the Jewish Religion. — The Progress of the Prophets reflected in their Conception of the Character of God. — Isaiah's God no longer the God of Israel alone.

HAVING now briefly adverted to a series of psychic phenomena recorded in the Pentateuch, it remains to note the continuation of the same through the prophets who succeeded Moses. It is undeniable that the phenomena exhibited by the prophets were the same as those of Moses in all essential particulars. They were dominated by the same beliefs, or suggestions, and the resultant manifestations necessarily corresponded, modified only by their different environment and the natural development and progress of the human mind. That the prophets were psychics, is undeniable. Even the methods sometimes employed by them in entering the psychical condition were identical with those often required by the modern psychic when preparing for some signal demonstration.

A striking illustration of this fact is found in the séance which the kings of Israel, Judah, and Edom had with the prophet Elisha. The King of Moab, who had formerly been tributary to the King of Israel, had rebelled. The latter had formed an alliance, offensive and defensive, with the Kings of Judah and Edom for the purpose of bringing the King of Moab to terms. On their march through the wilderness of Edom towards the land of the Moabites, they found themselves in a region that was devoid of water wherewith to water their stock. In this condition it became evident to them that they would soon fall an easy prey to the King of Moab. In this strait "the King of Israel said, Alas! that the Lord hath called these three kings together, to deliver them into the hand of Moab."

"But Jehoshaphat said, Is there not here a prophet of the Lord, that we may inquire of the Lord by him? And one of the king of Israel's servants answered and said, Here is Elisha, the son of Shaphat, which poured water on the hands of Elijah."[1]

Accordingly the three kings sought out Elisha, who, after making some disparaging remarks concerning the King of Israel, consented to do as requested for the sake of Jehoshaphat, the King of Judah.

"But now," continued the prophet, "bring me a minstrel. And it came to pass, when the minstrel played, that the hand of the Lord came upon him.

"And he said, Thus saith the Lord, Make this valley full of ditches."[2]

The point to be noted in the foregoing is that, on this great occasion, when the fate of three kingdoms trembled in the balance, Elisha deemed it essential that he should have the aid of music to enable him to enter the subjective state and successfully invoke the name of the Lord. It is needless to remark that it was precisely the condition often

[1] 2 Kings iii. 10, 11. [2] Op. cit., v. 15, 16.

required by the modern psychic to enable him to enter into communication with spirits of the dead for the purpose of obtaining their advice in cases of emergency.

It is also worthy of note that the "control" of Elisha recommended a very common-sense plan for obtaining water, namely, digging for it, since the "probabilities" did not promise rain in time to relieve the distress of the three armies.

What follows is illustrative of the essentially human and emotional character of Elisha's God, —

"And this [finding water] is but a little thing in the sight of the Lord; he will deliver the Moabites also into your hand.

"And ye shall smite every fenced city, and every choice city, and shall fell every good tree, and stop all wells of water, and mar every good piece of land with stones."[1]

And it was so.

Now, the prophets of Israel were undoubtedly the best men of that race. They it was who constantly enforced the monotheistic idea, and thus saved Israel from lapsing into idolatry. The hereditary priesthood represented the religion of the time in its external forms and ordinances. They were the guardians of its organization and its ritual. The predilection which the people evinced for ritual and ceremonial worship often betrayed them into acts of idolatry; that is to say, into the worship of other gods besides Jehovah. It was the prophets alone who constantly resisted this tendency towards polytheism on the part of the priesthood and the people. This conflict was carried on in a more or less pronounced form from the time of Moses to the time of Jesus; and to the successful resistance of the prophets to the insidious inroads of polytheism into the religion of the Jewish nation, is due the final triumph of Christianity.

[1] Op. cit., v. 18, 19.

It will thus be seen that *it is to psychic phenomena that the world owes its first conception of a living God.* The fact that the phenomena were grossly misinterpreted does not militate against the truth of this proposition. Its very nature was necessarily conducive to monotheism. Originating in the subjective mind of the psychic, it was inevitable that it should develop the emotional characteristics peculiar to the subjective mind.[1] One of the most pronounced of these characteristics, when it is not under the intelligent control of the objective mind, is that of monumental egotism. This emotion is developed in a more or less pronounced form in every phenomenal manifestation of subjective activity. The inevitable result was that, when once the idea was suggested that the source of the communications which were received by the psychics of the Mosaic Age was none other than the Deity himself, the character of the communications corresponded exactly to the psychic's conception of the character of God. That the first assumption should be that it was the greatest and most powerful of all the gods, was inevitable, especially when the psychic was imbued with the idea of a plurality of gods. The Jews, in common with all surrounding nations and peoples, were imbued with that idea. Idolatry and polytheism were everywhere dominant. The gods of other nations, however, were purely objective conceptions, and were represented by material objects of worship.

The God of the Jews, on the other hand, was evolved from the subjective intelligence of the seer by means of psychic manifestations. Just how the suggestion originated that the intelligence manifested in the phenomena was from God, must forever remain in obscurity; nor is the question of any great importance. That it originated with the Jews at a very early period of their national existence, is sufficiently

[1] For a full discussion of the distinctive characteristics of the subjective mind, see "The Law of Psychic Phenomena."

evident. That the great characters of early Jewish history actually existed, and that the story of their lives as related in the Old Testament is substantially correct, may, provisionally at least, be taken for granted. The important matter to be considered is not the physical history of the Jewish tribes, but the history of the evolution of the monotheistic idea. In the latter, however, we may find internal evidence of the substantial verity of the former. Certain it is that in the God of Abraham and of Isaac and of Jacob we find the first crude conception of the idea of a living God, — an idea which gradually grew in definiteness and magnitude until it was perfected by Jesus of Nazareth.

The God of Abraham, Isaac, and Jacob was the God of the Israelitish nation. The right of other gods to rule the other nations was not at first seriously disputed. It is true that the proprietorship of the whole earth was early asserted; but the whole attention of the Jewish God was occupied in providing for the wants and promoting the welfare of his chosen people.

This, then, was the first great step in the evolution of the spiritual man.

The point to be observed in this connection, and which must be constantly borne in mind, is that this step in the evolution of the spiritual man was brought about by means of psychic phenomena, — the phenomena of the soul. In the very nature of things it could be brought about in no other way. It is an axiom of the Christian Church that spiritual truth can only be apprehended through spiritual faculties. This is true in the sense that it is only by an observation of spiritual phenomena that a knowledge of spiritual truth can be obtained. It is obviously not true in the sense in which it is generally understood, — namely, that spiritual truth must always be perceived by the intuitional faculties of the soul; for the spiritual intuitions or perceptions of no two human beings were ever exactly alike, and they often differ as widely as the poles. It is obvious, therefore, that where the spiritual perceptions of two or more persons do not arrive at the same result, one or more of them has failed to perceive the truth. It follows that no intuition of the ordinary human mind can be relied upon as a guide to spiritual truth unless the intuitions of the great bulk of intelligent beings point to the same conclusion. History furnishes but a single instance of a man so exceptionally endowed with the faculty of spiritual or intuitional perception of spiritual truth that his teachings serve as a safe guide, not only to his own era and people, but to all mankind for all time.

Nevertheless, the intuitions of others cannot safely be ignored, for they often constitute important factors in particular cases, and frequently lead to a partial knowledge of great truths. Thus it is highly probable that intuition may have played an important part in the minds of the early prophets. It may have been an intuitional thought that led them to identify the intelligence manifested through psychic conditions with the living God. If so, it was a spiritual perception of a partial truth; for the soul of man is demonstrably a spark of the divine intelligence. Moreover, it is the divine instrumentality through which God manifests his will, and by which alone can his existence be demonstrated.

The next great step in the evolution of spiritual man was taken under the dispensation of Moses. He inherited the controlling ideas of his forefathers, and was consequently dominated by the suggestions embraced in the traditions of his race. Hence the God of Moses was the same anthropopathical deity that was worshipped by Abraham, Isaac, and Jacob. Nevertheless, a distinct and clearly defined step in advance was taken under the Mosaic dispensation. The code of ethics laid down in the Decalogue, rudimentary as it was, compared with the high standard of a later civilization, was a vast improvement on that of pre-Mosaic times. That this was largely due to the Egyptian education of Moses, there can be no doubt.

"If it be a fact, as there is no reason to doubt, that Moses was skilled in all the wisdom, esoteric and exoteric, of the Egyptians, there can be no difficulty in conjecturing the source from which he derived the code of rudimentary ethics which is laid down in the Decalogue. The scrolls and inscriptions which, in recent times, have been brought to light and deciphered, have demonstrated that long before the time of Moses the moral standard, theoretically at least, was very high in Egypt, as high indeed as that of the Decalogue. The great distinction of the Israelites — a very great one — was that their morality, even if it dated from their residence in Egypt, had the effect of soon refining and exalting their religious ideas, as was never the case in Egypt itself, where, curiously and inexplicably enough, a debased form of popular religion retained its place side by side with a high development, in some quarters or classes, of the moral sentiment." [1]

This "inexplicable" difference in the reciprocal effect of religion and morality in the two nations is easily accounted for when we take into consideration the difference in the two religions. A refined code of ethics and morality has no necessary effect upon an idolatrous form of worship; that is to say, they have no, or comparatively limited, reciprocal relations. On the other hand, there was a necessary and vital relation between the morality and the religion of the Jews. The God of the Israelites, as I have pointed out, resided in their seers and prophets. Their God could be communicated with, consulted and questioned; and the responses were the reflections of the mind of the one who consulted him. In other words, the God of the Israelites was just what they made him. If the prophet was dominated by a vicious code of morals, the responses from his God would be correspondingly vicious. If he was actuated

[1] The Natural History of the Christian Religion, by William Mackintosh, M. A., D. D.

by lofty sentiments and a pure morality, the responses from his God would be correspondingly elevated.

Hence it was that the religion of the Jews under the Mosaic dispensation was comparatively refined and elevated. Moses was their prophet and their seer, as well as their temporal leader; and, as Mackintosh remarks, —

> "The great fame and reputation of the Hebrew legislator is sufficiently justified by the fact that he so clearly discerned the importance of ethical and religious principles as a means of giving stability to social organization; that he took the highest results of the most ancient civilization which the world has seen, and laid them at the foundation of his nascent state; that he snatched the torch of human progress from hands that could bear it no further, and passed it on to those of a fresh and youthful race, — of a race which he may have freshened and rejuvenated by this very stroke of high policy." [1]

It seems probable, however, that the results were due, perhaps, less to high policy on the part of the great lawgiver than to the fortuitous commingling of the high code of ethics, which was a part of his Egyptian education, with the peculiar religion of his fathers. Be this as it may, the fact remains that the religion of the Jews, as it was in the beginning, — as it was when the first psychic seer conceived himself to be in communication with the Deity, — contained the germs of an infinite development; for every advancement in civilization, every step toward a higher grade of ethics, every new conception of a nobler code of morals, had a positive, direct, and necessarily corresponding effect upon those psychic manifestations which they believed to be communications proceeding directly from the living God. In scientific phrase, the suggestions embraced in their moral code were reflected and reproduced in their psychic manifestations, precisely as the so-called spirit communications of to-day are reproductions of the medium's

[1] Op. cit., p. 92.

conception of the character of the spirit who is supposed to be present.

The religion of the Jews was, therefore, necessarily elevated and refined just in accordance with the moral status of their prophets. But their prophets were human. They were subject to all the weaknesses and frailties of common humanity. They were restricted in their progress by the limitations of their environment; and civilization was yet in its infancy. Progress with them was, therefore, necessarily slow and fitful, and was often retarded by those reactionary forces which are always present in every inchoate nation.

It was many generations after the death of Moses before any substantial progress was made in the religious evolution of the Jewish people. Their prophets, however, continued to exist, and their influence gradually extended. The source of their inspiration was always the same; for the same suggestions were transmitted from generation to generation. Each prophet believed himself to be in communication with the only living and true God. The separation of the tribes after the death of Moses did not change the dominant idea. Each prophet of all the tribes believed himself to be in communication with the only true God. Each tribe regarded with contempt the gods of the other tribes, as was natural. Nevertheless the monotheistic idea was dominant with all the prophets. It could not be otherwise; for the inherited suggestion was constantly before each of them that when he entered the psychic state it was because "the hand of the Lord was upon him;" and his utterances when in that state were necessarily in keeping with the dominant suggestion that he was giving voice to the very words of the living God. Hence the authoritative prefix to all their announcements — "Thus saith the Lord" — was uttered in the solemn tones of sincerest conviction.

That they were earnest and sincere in their convictions,

cannot be doubted. Their enthusiasm, which oftentimes developed into fanaticism, could be the result of nothing less than absolute conviction, not only of the truth of their utterances, but of their divine origin and authority. They could not believe otherwise; for they were constantly experiencing phenomena which forced that conviction upon them. They found themselves habitually entering a state that was to them mysterious and abnormal, and yet agreeable. In that state they entered into communion with what appeared to them to be an extraneous intelligence. That intelligence sometimes put words into their mouths that were foreign to their objective thoughts. In short, they experienced the same phenomena that modern psychics attribute to disembodied spirits, differing only in the suggestions which gave character to the manifestations. To them the evidences of divine communion and of their own divine mission were demonstrative. That conviction was communicated to the people, partly by their intense earnestness, and partly by their occasional exhibitions of psychic power in the performance of what was then regarded as miracles, as in the cases of Elijah and Elisha and later by Jesus of Nazareth.

It will thus be seen that the monotheistic idea was inherent in the very nature of the psychic phenomena experienced by the seers and prophets of the Jewish race. The first psychic who, no matter how the idea originated, conceived himself to be in communion with God, fixed the monotheistic idea in his own mind and in the minds of his successors for all time. No power on earth could uproot that idea thus formed, so long as there was a succession of psychics to whom the dominating suggestion could be transmitted. Nevertheless it was an idea that possessed the seeds of future development, in that every step in the progress of a higher civilization correspondingly elevated and refined the popular conception of the Deity. We have already seen the advancement of the idea under the dispensation of Moses. Its development after his death was comparatively slow up to the advent of Jesus. Nevertheless, there was substantial progress before that event. The God of Abraham was the God of a tribe. As the tribe enlarged to a nation, he became a national God. But it was not until the days of Isaiah that the idea of God's power and dominion was so enlarged as to give him credit for benevolent intentions towards the whole race of mankind.

"And he said, It is a light thing that thou shouldst be my servant to raise up the tribes of Jacob, and to restore the preserved of Israel: I will also give thee for a light to the Gentiles, that thou mayest be my salvation unto the end of the earth." [1]

In chapter lx. of Isaiah, there is further vague mention of the benevolent intentions of the God of Israel towards the Gentiles; but it is also evident that the latter were of secondary consideration: —

"Yet it must be confessed," says Mackintosh, "that even the prophetic hold of this higher conception was wavering and unsteady, as is conspicuously apparent in the psalter, where the old popular, or we may say heathenish, and prophetic sentiments follow each other in baffling confusion, in irreconcilable juxtaposition. Not a reader but is surprised, if not pained, to see the breath of vengeance and the breath of mercy blow by turns through those wonderful compositions, which were probably among the last, and were, in some respects, the greatest products of the prophetic spirit."

The most substantial progress of the monotheistic idea during the interval between the death of Moses and the advent of Christ, consisted not so much in the enlargement of the idea itself as in its general acceptance by the Jewish people. This was brought about by their misfortunes, together with the earnest efforts made by their

[1] Isaiah xlix. 6.

prophets to convince them that it was to their wickedness, evinced in following after false gods, that their troubles were due. With the details of their history, however, we have little to do. It suffices to know that the monotheistic principle was developed by and through the Jewish people by means of psychic phenomena. As I have before remarked, the fact that it was misunderstood at the time does not militate against the broad truth of the proposition. All the phenomena of Nature were misinterpreted in the infancy of the human race; but the laws of Nature were the same then as they are now. The advantage that we possess consists in the fact that we know a little more than did the ancients of the laws which pertain to the phenomena of Nature. They thought that the earth was flat, and that it was the centre of the universe. We know that it is round, and that it constitutes but an infinitesimal part of the universe. We also know that it was round when they lived, and that it will continue to be round as long as it holds together. In other words, with our knowledge of astronomical laws we can reconstruct the past and foretell the future movements of the heavenly bodies. We also know something of the laws of the soul. It is little, it is true, compared with what is known of the physical sciences; for the fact is just fairly beginning to dawn upon the human mind that the soul can be studied scientifically. But enough is known of the phenomena of the soul to enable us to classify a few of the leading facts and understand something of their significance. The phenomena which we have been discussing have a well-settled place in psychic science, and there is no longer any excuse for misinterpreting them.

CHAPTER VIII.

THE ADVENT OF JESUS.

The Third Great Step in the Evolution of the Spiritual Man. — The God which Jesus Proclaimed. — Intellectual Prodigies. — The Intuitional Powers of Jesus. — His Psychical Powers. — His Perfect Knowledge of the Laws of the Soul. — Modern Confirmations of the Truth of his Philosophy. — The Psychic Methods of Jesus. — His Reason always in the Ascendant. — His Perfect Moral and Religious Character. — Psychic Phenomena the Evidence of his Divine Mission. — Paley's Views. — The Divine Heritage. — The Vitality of the Christian Religion.

THE third great epoch in the evolution of the spiritual man was inaugurated by Jesus of Nazareth. In discussing this branch of the subject, I shall not enter the field of controversial argument respecting his alleged miraculous birth or his resurrection from the dead. I leave that to the theologian who regards those questions as possessing vital importance from his point of view. From my standpoint they cannot be considered. Miracles can have no place in science; for they can neither be scientifically verified nor experimentally reproduced. I have thus far confined my observations to the records of such psychic phenomena as can be verified by experimental reproduction; and I shall continue to do so wherever the nature of the phenomena will admit of such demonstration. In the history of Jesus, however, there is much that cannot be specifically verified by experiment. His character and inherent attributes cannot be reproduced. We are not, however, without means of scientifically verifying his history in that regard, as will be seen later in this work.

The phenomena which we are now called upon to consider differ in many essential particulars from those recorded in the Old Testament. The older prophets, as we have seen, were psychics who believed that they had the power to enter at will into tangible communion with God, and to receive from Him direct verbal communications. These phenomena, as I have pointed out, were identical with the phenomena of modern spiritism, differing only in the suggestion which gave character to the supposed communications. The God of the old prophets was, therefore, necessarily a reflection of their own personal characteristics. On the other hand, the God whom Jesus revealed to mankind was a conception so grand and lofty, as compared with that of the old prophets, that credulity has, in all the ages, been taxed in vain to identify the God of Abraham with the God of Jesus. This fact has been a stumbling-block to the sceptic or heretic for eighteen hundred years; whereas, when the facts are understood, they will be found to present the strongest possible internal evidence of the substantial truth of the essential portions of the historical part of both the Old and the New Testaments. Viewed as a series of psychic manifestations, the gradual improvement in the God of Israel corresponds exactly with the natural progress of that race towards civilization, and the consequent evolution of the human mind and soul. That there was a sudden step in advance, of infinite magnitude and importance, does not militate against the theory of the evolution of the spiritual man through psychic phenomena. On the contrary, it will be found to confirm and emphasize that hypothesis. The great step in advance which Jesus made was the result, not of a cessation of psychic manifestations, but of a radical change in their character. The conception of God which he evolved was not the result of verbal communications from

God, after the manner of the prophets, but was the result of the fact that *he was endowed with the faculty of intuitive perception of the laws of the human soul.*

In order to understand fully the position which Jesus occupied in the psycho-religious world, it will be necessary briefly to discuss the above proposition. In "The Law of Pyschic Phenomena" I have discussed it at some length, and space can be given here for but a brief outline.

History shows that from time to time there have been born into this world persons so exceptionally endowed by nature with intellectual powers in certain directions that they can be appropriately designated by no other term than that of "prodigies." The phenomenal manifestations of these prodigies, however, are usually confined to some one sphere of mental activity. Thus, there are musical prodigies, prodigies in art, in poetry, in mathematics, etc., but seldom has one of them been known to be exceptionally endowed in more than one direction. The salient characteristic which is common to all is that each one appears to be endowed with an intuitive perception of the laws of nature which pertain to his specialty. Thus, the musical prodigy intuitively perceives the laws pertaining to the harmony of sounds; the artistic prodigy intuitively perceives the laws of harmony of colors; the mathematical prodigy perceives by intuition the laws of numbers. This faculty when once developed absolutely transcends reason or objective education. Neither the one nor the other has any part or lot in the production of the phenomenal results. Fortunately we have the means of verifying this proposition. Thus, one of the most phenomenal musical prodigies the world has ever known was both blind and idiotic from birth.[1] Obviously, therefore, neither objective reason nor objective education could have played any part in his musical development. Yet he was able when a mere child to improvise

[1] Blind Tom.

excellent harmonies, and to reproduce a piece of music, once heard, with remarkable accuracy. Thus the proposition is scientifically verified that he had the power of perception of the laws of nature which governed his specialty; for there is no other way of accounting for the phenomena.

Again, the mathematical prodigy, Zerah Colburn, mentioned at length in "The Law of Psychic Phenomena,"[1] before he was objectively able to understand the powers of the nine digits could instantaneously solve intricate arithmetical problems. He was investigated and scientifically tested by the ablest scientists of Europe, who bear testimony to his prodigious powers. His answers were given so promptly that calculation was out of the question, even if he had been educated in the rules of arithmetic, which was not the case. Again we have a scientific verification of the facts related of him, and of the proposition that his powers were the result of intuition, in this, that (1) he developed his powers before he had studied arithmetic; (2) his answers were instantaneously given; (3) his answers were always correct. The last was, of course, the supreme test, for the reason that if they had not been correct they would not have been remarkable.

Many other mathematical prodigies might be mentioned in this connection did space permit. Their phenomena are no more remarkable than those of prodigies in other sciences; but they are more valuable for purposes of generalization than any others, for the reason that they carry with them their own verification. Their answers, being accurate, demonstrate their powers of intuition; and they also demonstrate the general proposition that the soul of man possesses the inherent power, under certain psychic conditions not yet clearly understood, to perceive by intuition the laws of Nature.

This proposition conceded, it is easy to account for the

[1] See also "Memoir of Zerah Colburn."

knowledge which Jesus possessed of those laws of Nature which pertain to the functions, powers, and destiny of the human soul. He had an intuitive perception of those laws, and his knowledge of them was undoubtedly as accurate as if it had been susceptible of mathematical verification.

At this point it will be asked: "What evidence have we that Jesus was endowed with that power of intuition?" This is a most pertinent question, and, could it not be clearly and logically answered, we should be compelled at this point to abandon all scientific methods of inquiry into this subject. Fortunately the proofs are at hand which will settle that question beyond all possibility of reasonable doubt.

In the first place, it must be remembered that Jesus was born and reared among a people who believed in and worshipped the God of Abraham and of Moses. Their ideas of God were based upon the purely anthropomorphic conceptions of the Deity which had dominated the race from Abraham down. His education, therefore, other things being equal, was calculated to inspire him with the beliefs of his ancestors. But other things were not equal. He was exceptionally endowed, morally, as he was intellectually and psychically. He was infinitely above his race in every attribute which contributes towards human perfection. He may or he may not have entered the psychic state in order to hold communion with God, as did the prophets before him. It does not seem probable that he so acted, for there is nothing in his history that points to that conclusion. At the age of twelve he was able to dispute with the doctors in the Temple in his normal condition. But, even if he had entered the psychic condition for such a purpose, his conception of God would have been infinitely above that of the older prophets, and would still have constituted a mighty step in the evolution of spiritual humanity.

It is to the last degree improbable, however, that he ever entered the psychic state with the idea of receiving verbal communications from God. His conception of God was far too lofty for him to be led into the errors of his predecessors. His intuitive knowledge of the laws of the soul would of necessity prevent him from placing himself in a position where he could possibly be dominated by a false suggestion. The whole history of his life shows that he never allowed his subjective mind to usurp the throne of his reason. His Sermon on the Mount demonstrates his entire emancipation from the thraldom of precedent, and proclaims, inferentially, his realization of the finite character of the God of Moses. By his frequent repetition of the words, "Ye have heard it said by them of old time," etc., followed by his "But I say unto you," etc., he placed in violent contrast the God of Moses and his own conception of the Deity, and of the duty of man towards his Creator and his fellow-men. He could have used no language that would have more utterly repudiated the Mosaic conception of the God who gave to Israel the imperfect code of ethics delivered to Moses on Mount Sinai. That he did not proclaim, in so many words, his knowledge of the human origin of the God "of old time," was doubtless due to that caution which he so often evinced in speaking to the people in parables, and which was expressly stated to his disciples in the following memorable and significant words: "I have yet many things to say unto you, but ye cannot bear them now." [1]

He gave to the people what they could readily assimilate, and he always refrained from unnecessarily antagonizing their ancient beliefs and prejudices.

One of the strongest evidences of the intuitive character of his knowledge is found in his conception of the character and attributes of God. This, in connection with the

[1] John xvi. 12.

fact that he was the first to proclaim a conception of the Deity so lofty, so grand, and so ennobling, and at the same time in such perfect harmony with the highest instincts of all civilized humanity, constitutes a strong link in the chain of evidence to sustain that hypothesis. Born of a race whose highest conception of a Deity was of a being whose passions and weaknesses would degrade a savage, and whose highest purpose it was to protect a single tribe or race in no wise better than their neighbors, Jesus proclaimed a God of love, mercy, and benevolence, and promulgated a code of ethics for the guidance of the human race, the fundamental principles of which were the universal brotherhood of man and the Fatherhood of God. That his code of ethics and morals, and his conceptions of the attributes of God, have never been and never can be improved upon, no one will undertake to deny. That they constitute strong evidence, not only that he was endowed with an intuitive perception of the laws pertaining to the subject-matter, but that his intuitions were correct, is evidenced by his undisputed headship and by ample time-tests.

The strongest evidence, however, of the fact that he possessed the power of intuitive perception of the laws of the soul, consists in his physical manifestations. It is true that his miracles belong to a comparatively low order of psychic phenomena; but it was absolutely necessary for him to display his powers in some tangible form in order to impress his followers with a sense of his power and authority. "Except ye see signs and wonders, ye will not believe," was a statement, made to the nobleman of Capernaum, of a pregnant fact.[1] It was a recognition of an existing condition of the public mind with reference to him and his claim to divine authority. It was a proclamation of his purpose

[1] The learned Nicodemus (John iii. 1) says: "We know that thou art a teacher come from God: for no man can do these miracles that thou doest except God be with him."

to satisfy the public demand; and his wisdom was never more manifest than in his compliance with the popular desire to witness exhibitions of his power. For he not only gave to them the only proofs of his divine mission that they could appreciate, but by the same means he left a record of his works which now constitutes the only means we have of verifying his history.

It is a singular fact in the history of the Christian religion that the circumstances and events in the life of Christ which have been the greatest stumbling-blocks of scientific scepticism for eighteen centuries, are, in this last quarter of the nineteenth century, found to be the only facts in his history which can be scientifically verified. The most potent assaults of scepticism have been made upon the record of his physical manifestations. Thousands who could have accepted without serious question the fact of his spiritual supremacy, who admired his code of morals and reverenced his exalted character, have derisively rejected the story of his miracles and ended in total scepticism. In a scientific age this was inevitable. The moment one begins to comprehend the principles of induction, the moment one realizes the constancy of the forces of Nature and the immutability of her laws, that moment the seeds of scepticism are implanted in his mind, and miracles are relegated, in his philosophy, to the domain of fable or of superstition. The Church, in turn, has provoked this spirit of scepticism by constant iteration of the dogma that Christ wrought his wondrous works outside, and in defiance, of natural law. Fortunately for the Christian Church and for humanity, the scientific investigations of the last quarter of the nineteenth century have revealed the fact that the so-called miracles of Christ can be experimentally reproduced. Moreover, the laws which governed the production of his phenomena are beginning to be understood; and some of the more important of them have been definitely formulated, and have been incorporated into the great body of modern science. Since that has been accomplished, it is suddenly recollected that Jesus himself never claimed to perform his works outside of natural law. On the contrary, he not only taught his Apostles how to reproduce his phenomena, but proclaimed to the world the essential conditions to their reproduction, and declared in so many words that those who observed those conditions should be able to do "even greater works" than he had done. Modern science has rediscovered the art of doing those works; and it has formulated the conditions necessary to be observed. And it is just here, therefore, that the most positive evidence of the essential truth of the history of Jesus of Nazareth is to be found. Considering his physical manifestations as miracles, science must forever discredit his history. But when it is discovered that in the nineteenth century the lame can be made to walk, the blind to see, and the deaf to hear, just as he did those things in the first century, science has no more right to discredit his history than it would have to dispute any other historical instance where the forces of Nature had been utilized; *a fortiori*, where there was indubitable evidence of their intelligent utilization. This evidence we have in the history of Jesus, in that *he minutely observed all the conditions that modern science has discovered to be necessary for the successful reproduction of his phenomena.*[1]

It will thus be seen that the Christian religion forms no exception to the rule that all religions worthy of the name have their origin in psychic phenomena.

Jesus himself was the most stupendous psychic phenomenon the world has ever seen. He was a colossal religious genius. Endowed with a perfect power of perception of all the laws of the human soul, he was enabled to formulate, and to promulgate to the world, a series of vital truths and

[1] For a full discussion of these points, the reader is referred to "The Law of Psychic Phenomena."

principles which the most profound researches of inductive science can only verify.

I have elsewhere defined "genius" as the result of the synchronous action of the objective and subjective faculties.[1] History records the names and works of many men who have merited this designation to a certain extent. A few have given evidence that, in a purely intellectual sense, the synchronism was well-nigh perfect. Many have given occasional exhibitions of intellectual power which can be accounted for on no other hypothesis. History, however, furnishes us but one example of a man in whom the synchronism of development, physical, intellectual, psychical, and moral, was absolutely perfect. That man, it is needless to say, was Jesus of Nazareth. That Jesus was a psychic of most wonderful power, no one will gainsay, whatever may be his theories regarding his miraculous conception and birth; but he was, in many essential particulars, unlike any other psychic of whom we have any record.

The ordinary psychic, in order to produce his phenomena, is compelled to enter the psychical or subjective condition. His objective faculties must be and remain in at least partial abeyance. In this condition his objective reason is dethroned and he is dominated by the power of suggestion. His phenomena will, therefore, necessarily take the form of whatever suggestion is uppermost in his mind; whether it be an auto-suggestion arising from his preconceived opinions, as in spiritistic phenomena, or a suggestion from another, as in hypnotism. In any event, his objective reason is in abeyance, and consequently, if the suggestion is a false one, he is nevertheless dominated by it, and the resultant phenomena are necessarily incongruous and misleading. It is true that in the phonomena which are the products of what is known as "genius," there often appears to exist a perfect synchronism of objective and subjective activity and

[1] See "The Law of Psychic Phenomena," ch. v.

development. This, however, pertains only to purely intellectual manifestations; and it is rarely, if ever, constant. But where this synchronism exists it has never been known, in modern times, to be accompanied by the power to produce other psychic phenomena, especially physical manifestations. The latter, as before stated, are only produced, by the ordinary psychic, as a result of his entering the subjective state, in which the objective faculties are held in partial or complete abeyance.

In many of the foregoing particulars Jesus constituted an exception to the general rule. Not that he was exempt from the operation of the universal law governing psychic manifestations, but that he was, to a greater extent than any other psychic, harmoniously developed. In him the objective and subjective faculties preserved, at all times and under all circumstances, an exquisitely harmonious balance. Having an intuitive perception of psychic laws, he was fully aware of the ill effects of their misdirected application. Knowing the limitations of the powers of the subjective mind, its amenability to control by suggestion, and its consequent inability to take the initiatory step in the process of induction, he never allowed it to obtain control of the dual mental organization. Consequently, his reason was always in the ascendant; and history does not record an instance where he entered the psychic state for any purpose whatever. Moreover, he never allowed himself to produce any phenomena for the mere purpose of displaying his powers. When he consented to exercise the powers of the soul, it was always for the purpose of accomplishing some good object; albeit his primary object may have been to convince the people of his divine mission. Be this as it may, the fact remains that he never exercised his psychic powers except for the promotion of the highest good of those around him; and he never allowed himself to be placed in such a mental condition as to render it possible for him to be dominated by a false suggestion.

It will thus be seen that his wisdom was as strikingly displayed by what he refrained from doing as it was by what he did. Indeed, it will eventually be found, as knowledge of psychic laws increases, that one of the most valuable lessons which Jesus taught to mankind consisted in his abstention from any unnecessary display of his psychic powers. Knowing, as he did, the laws pertaining to the production of psychic phenomena, he carefully and consistently kept within the normal lines. The story of his three temptations in the wilderness was nothing more nor less than a symbolical presentation of this most important principle by which he was guided. Rightly interpreted, the story of the three temptations draws the line of demarcation clearly and distinctly between the legitimate and the illegitimate — the normal and the abnormal — exercise of psychic power.[1]

Formally stated, the distinctive characteristics of the psychic powers and attributes of Jesus are as follows: —

1. The first and most important mental characteristic which distinguished him from all other men of whom history has given any account, consisted in his intuitive perception of the laws which pertain to the human soul. I say that this was his most important mental endowment, for the reason that it was the essential prerequisite to all the others. It not only enabled him to "speak as never man spake," that is, with the authority of perfect knowledge; but it enabled him to exercise the powers of the soul under the most favorable conditions.

2. The next in importance of his distinctive endowments was his ability to exercise his psychic powers under normal physical conditions. No other psychic has ever been able to do this except to a very limited extent. This ability arose, not because he acted outside the

[1] For a fuller discussion of this proposition, see "The Law of Psychic Phenomena," ch. xxiv.

domain of natural law, but from his perfect knowledge of the law. Other psychics have sometimes performed purely intellectual feats while in an apparently normal physical condition. Some have, under exceptionally favorable circumstances, produced physical manifestations under apparently normal conditions. But such cases are sporadic, and only serve to emphasize the general rule that psychic manifestations are the result of abnormal physical conditions. Jesus was the only one, of whom we have any authenticated account, who never found it necessary to enter the subjective state to enable him to produce any psychic phenomena.

3. As a result of his ability to exercise his psychic powers without entering the subjective state, he was enabled to avoid the operation of the law of suggestion, and, as a consequence, he was never dominated by a false suggestion. Objective reason, therefore, was always in the ascendant. Again it must be remarked that this does not imply that he was not subject to the law of suggestion; but that his knowledge of the law enabled him to avoid placing himself in that condition in which he would be dominated by it. It is in psychic science as in any other. If we know its laws we can avoid its evils. In other words, when we are dealing with a force, of the laws of which we have perfect knowledge, we are enabled to place ourselves in proper relations to it, and thus avoid the penalties attending the infraction of its laws.

4. One of the most important of the distinctive characteristics of Jesus, as compared with other psychics, consisted in his perfect moral and religious character. This, in a certain sense, may be attributed to a perfect knowledge of the laws of the soul; although due credit must be given to that innate altruism which was regnant in his character. Without attempting, however, to distinguish

between what was the result of a perfect comprehension of spiritual laws and what was innate in his character, if indeed there is any line of distinction, it will be sufficient for our present purpose to discuss the former. As I have before remarked, the subjective mind or soul is the seat of the emotions. Every emotion, therefore, is a psychic phenomenon. Religious worship is an emotion that is inherent in every human soul. It is one of the higher instincts which differentiate the man from the brute. Morality is also an emotion when considered as a duty which man owes to his Creator, although, when practised solely with reference to one's relations to society and the commonwealth, it is the result of education. In its highest sense, therefore, morality is an emotion, cognate to religion, and, with the latter, must be considered as one of the phenomena of the soul. Religion and morality being phenomena or attributes of the soul, they necessarily have normal relations to every other attribute of the soul. This being granted, it follows that one who possesses a perfect knowledge of the laws of the soul will be able to discern those normal relations, and, other things being equal, will seek to maintain them. This, then, was the distinctive characteristic of Jesus. He was a master of the science of the soul, and as such had a perfect knowledge of its attributes and powers, and of the normal relations which those attributes and powers sustain to each other, to humanity, and to the Creator. A perfect moral and religious character was necessarily the result.

It will thus be seen that, as before remarked, the Christian religion forms no exception to the general rule that every religion worthy of the name has its origin in psychic phenomena. Previous to the time of Jesus, the phenomena were grossly misinterpreted. Nevertheless, they contained the germs of the monotheistic idea, which was perfected by one who never misinterpreted the phenomena of the soul,

Jesus of Nazareth. The Christian religion not only had its origin in psychic phenomena, but that was the only means by which it was, or could be, brought to the knowledge of mankind. The words of Jesus would have been lost, and his mission a· failure, had he not been endowed with the power to produce phenomena tangible to the senses of the people. It was by this means alone that he was able to impress upon the world a realization of this divine mission. Paley himself declares this fact in these words: —

"That this particular person, Jesus of Nazareth, ought to be received as the Messiah, or as a messenger from God, they [the Apostles] neither had, nor could have, anything but miracles to stand upon."[1]

It was by the miracles that Jesus was enabled to impress upon his followers a sense of his power, and of his authority as a messenger of truth. And what was true of his immediate followers is largely true of the Church from that day to the present. It matters not that the so-called miracles were misinterpreted psychic phenomena. They were not misinterpreted by Jesus himself; for he never claimed that he performed his works outside the domain of natural law. On the contrary, he distinctly proclaimed the fact that others could do even greater things than he had done by complying with the conditions which he prescribed. It would have been idle for him to attempt to explain to his followers the underlying scientific principles which enabled him to produce his phenomena; for no one of his day was capable of comprehending them. Moreover, if he could have succeeded in convincing them that he did not transcend the laws of Nature in the performance of his miracles, it would have weakened their confidence in his divine mission; for the people of that day were incapable of grasping

[1] Evidences of Christianity, ch. x.

the idea that God could possibly display his power in any other way than by some signal violation of his own laws.

It will thus be seen that the Christian religion not only had its inception in psychic phenomena, but that faith in it has been perpetuated largely by a misapprehension of the real significance of the psychic manifestations of Jesus. If, however, the miracles alone had constituted the evidence of the truth of Christianity, it would have long since perished as a system of religion. For no system of religion which is founded upon a fundamental error can long withstand the assaults of scientific scepticism, in an enlightened age and nation, where truth is left free to combat error. In a primitive age a claim to supernatural power may serve to impose almost any system of religion upon a people. In an enlightened age such a claim is an element of weakness; and a theology founded upon that alone must eventually perish and be forgotten. The assaults of scepticism upon the Christian religion have been almost exclusively upon the dogma of supernaturalism; and had its claims to a divine origin rested alone upon that, it must have yielded to the first onslaught of scientific scepticism. That it has sustained the shock of scientific criticism, and is still a great and growing power in the most enlightened age the world has ever seen, and is now the most potential force in the social systems of the most enlightened nations of the earth, is indubitable evidence that it possesses an inherent vitality that must be looked for outside the domain of the supernatural. In the ensuing chapter I propose briefly to inquire into the secret of the wonderful vitality of the Christian religion.

CHAPTER IX.

THE INTUITIVE PERCEPTION OF TRUTH.

Books that thrill the Reader with Pleasurable Emotions. — Theories to account for it. — Literary Style. — Personal Magnetism. — The Soul's Love of Truth. — Books Popular in proportion to their Truth. — The Scriptures. — The Philosophy of Jesus. — Intuitional Perception of its Truth. — Evolution of Religion. — Christianity the Final Goal. — The Impossibility of improving upon True Christianity. — The Absolute Religion.

IT has often been remarked by intelligent readers of books that some authors have a faculty of impressing their personality upon their literary productions; so that one experiences, when reading them, a thrill of pleasure and satisfaction akin to that felt when listening to an orator who possesses what is known as great "personal magnetism." Some have attributed this feeling wholly to the literary style of the author; whilst others, more prone to suspect that an occult force is concealed behind every phenomenon, have held that the "personal magnetism" of every author is, in some inexplicable way, impressed upon the pages of his book. It seems obvious that neither of these explanations can possibly be the true one.

The first cannot be true, for the reason that it often happens that works which create the deepest impression upon mankind are written in a very unattractive style; whilst other works leave no lasting impression upon the minds of their readers, although couched in terms of faultless elegance. The second explanation is defective, even to absurdity; for whatever occult force, personal magnetism, or psychometric or telepathic impression might be supposed to accompany an author's personal manuscript, it is obvious that it could not be transmitted to the printed page which the author never saw or handled. Besides, it often happens that editions of an author's works are printed hundreds of years after he is dead; but it has never been noted that the element of so-called "personal magnetism" diminishes in force or intensity as the editions of his works are multiplied. The thrill of satisfaction which every man of intelligence feels when reading the lines of Shakespeare is not diminished in intensity as the years go by; nor does it suffer any appreciable change since it has been claimed that they were written by the "greatest, wisest, meanest of mankind." It is evident, therefore, that we must seek elsewhere than in elegance of diction or personal magnetism for an explanation of the secret of the permanent popularity of a book.

Broadly speaking, a book is permanently popular in proportion to the amount of truth it contains. Works of fiction constitute no exception to this rule; for our appreciation of a novel is in exact proportion to the fidelity to nature with which its characters are portrayed. What is true of a work of fiction is necessarily true of a work professing to deal with facts, as in history, or with principles, as in science, in philosophy, or in religion.

The love of truth is inherent in the normal human soul, and its recognition of truth is instinctive. This in itself constitutes a psychic phenomenon of the utmost importance; and it is one which must enter as a factor into every correct diagnosis of the attributes of the psychic entity. It is this instinctive perception or recognition of truth when it is presented that gives rise to that emotional thrill of pleasure and satisfaction which one experiences when reading the statement of a vital truth. It is the soul's response to a suggestion which is in accord with its own deductions from the facts of its own experience. In this connection it must be remembered that the memory of the subjective mind is perfect, and that its power of deductive reasoning is also perfect. It is, however, devoid of the power of induction proper, being constantly amenable to control by suggestion. When, therefore, a suggestion is imparted to it that corresponds to its own deductions, it instantly recognizes its truth and responds with a thrill of pleasurable emotion. This emotion alone is indubitable evidence that it is a purely subjective experience, since the subjective mind or soul is the seat of the emotions as well as the storehouse of memory.

This phenomenon is experienced in a greater or less degree upon the perusal of any book which contains what the reader recognizes as truth; and the intensity of the emotion experienced is in proportion to his estimate of the degree of importance to be attached to it as affecting himself. For the purpose of this inquiry, however, books must be divided into two general classes. Those which treat of temporal affairs belong to one class, and those which deal with questions pertaining to the attributes, powers, and destiny of the soul belong to the other. Those belonging to the first class never produce the phenomenon proper of which we speak. Such books may be never so interesting or important to the temporal well-being of man, yet they rarely, if ever, produce other than a purely intellectual enjoyment.

On the other hand, that which pertains to the soul is taken cognizance of by the soul, which is moved to emotion, pleasurable or otherwise, just in proportion to its recognition of the vital truths which a book contains. By this it is not meant to convey the implication that the emotions experienced on reading a book are infallible standards of truth. On the contrary, our subjective perception of truth is oftentimes neutralized by our objective perceptions or prejudices, or from those primordial anterior suggestions arising from fixed habits of thought or moral principles. But truth possesses an inherent vitality which no amount of error can wholly extinguish. In the long run truth must prevail, in spite of passion and prejudice. Hence it is that books which contain vital truths, however modest their pretensions or homely their style, will be enshrined and live forever in the hearts of their readers, whilst the more pretentious volume, devoid of the vitalizing element of truth, though adorned with all the perfections which learning and eloquence may impart, makes no permanent impression upon the souls of men, and is soon forgotten by the intellectual world.

The faculty of perceiving those truths which affect the human soul is inherent in the soul, although it is in rare cases only that it is largely developed in any one individual. Jesus was probably the only man who was endowed with this faculty in perfection; that is, he was the only one, of whose life we have any record, who possessed the power of independent perception of the laws of the soul. Others possess that power only in the limited sense that they are able to grasp and comprehend the truth when it is presented to them. But in that sense it is so generally diffused among mankind that in the aggregate it must be counted as a most important factor in the social, moral, and religious world; and in an enlightened community it prevents any radical misconception of the fundamental principles of morality and religion.

The intelligent reader will have anticipated me in what I am to say regarding the practical application of these observations to the fundamental principles of the Christian religion. It seems to me, that is to say, that the fact that Christianity still exists as a system of religion, is evidence, little short of demonstrative, that it is founded upon the true science of the human soul. It is certainly the strongest possible corroborative evidence of the truth of the claim that Jesus correctly expounded the laws of the soul in its relations to the divine intelligence. There can be no other rational explanation of the pregnant fact that the Christian religion has survived the assaults of its enemies for nearly nineteen hundred years, and is still the religion of the most enlightened nations of the earth. It has not only survived the assaults of its enemies, but it flourishes in spite of the mistakes of its friends. If it had not been founded upon the rock of Eternal Truth, it might have temporarily imbibed a vitalizing inspiration from the opposition of conflicting religions, but it never could have survived the proselyting methods of Charlemagne, the zeal of the Inquisition, or the dogma of plenary inspiration.

It is safe to say that no system of religion has ever flourished amidst so many adverse conditions as has the Christian religion. It had its roots in a region remote from the centres of civilization, and among a nomadic race, who were poor, and despised and reprobated and persecuted by their more powerful neighbors. From the first it encountered the refined philosophy of the most enlightened nations of the earth, and it has been engaged in stubborn conflict with all the material science of modern civilization. It has its literary setting in a volume which teaches an absurd astronomy, an impossible geography, and a cosmogony the crudeness of which is detected and exposed by the learning of every school-boy.

And yet it exists, not in decrepitude and decay, but as a vital element in every civilization worthy of the name. Its votaries have thrust it into conflict with every science, and it has been defeated in every encounter. Yet it is not relegated to the domain of ignorance, but flourishes in the greatest luxuriance of growth and vitality in those nations whose people are the most enlightened and progressive.

That there is to be found, within the realm of natural causes, some good and sufficient reason for this apparent paradox, is not to be doubted. The explanation afforded by the doctrine of a continuous miracle must be regarded as scientifically untenable. It seems to me that the following propositions afford at least a partial solution of the problem: —

1. Jesus Christ was endowed with the faculty of intuitional perception of the natural laws of the human soul; and he proclaimed to mankind, in a few simple propositions, the essential principles which govern the relationship of man to his fellow-man and to God.

2. All men are endowed with the same intuitional powers, differing only in degree; and by this means they are enabled to recognize, when once presented, any truth which is essential to the welfare of the human soul.

3. It follows that, when one reads the simple but all-comprehensive philosophy of Jesus, his soul intuitively and instantaneously recognizes its essential truth.

This is what has been, by the Church, vaguely denominated a "spiritual perception of religious truth," — a phrase which describes the emotion correctly enough, but which has never itself been scientifically or philosophically explained. When the emotion of religious worship, which is an inherent attribute of every normally developed human soul, is taken into consideration, it will be readily understood why it is that the Bible affords consolation to such a vast multitude of the human race. It is not alone the words of Jesus which proclaim religious truth, but scattered all through both the New Testament and the Old may be found passages innumerable upon which is stamped the sign-manual of eternal truth. Variable and diverse as are the emotions and aspirations, the spiritual wants and necessities of aggregate humanity, there may be found in the Scriptures something to fit every case, something to pour the balm of consolation into every stricken breast, something to inspire every human heart with hope; in short, in its power of adaptation to all the experiences of human consciousness, the Bible is unequalled by any other production, human or divine.

The philosophy of Jesus, however, constitutes the chief corner-stone of the whole superstructure. It is that which imparts vitality to the whole body of religious doctrine contained in the Bible, which but for that philosophy would have long since yielded to the assaults of scientific scepticism. But vital truth can never be wholly obliterated, however thickly it may be overlaid with error. It may be temporarily obscured, but the intuitive powers of the soul are safe guides to its recognition wherever found. Hence it is that the Christian religion has never lost its inherent vitality amidst the adverse influences with which it has been surrounded, but constitutes the essential vitalizing force in the civilization of every enlightened nation.

I do not undertake to say that these facts constitute conclusive proofs of the truth of the doctrines of Jesus; but, from a logical and scientific standpoint, it cannot be doubted that they constitute presumptive evidence that, in its essential features, his philosophy bears the impress of truth. I certainly know of no other way of accounting for the hold which the Christian religion has upon the mind and heart of civilized humanity, than to suppose that it is the aggregate result of the inherent power of man to recognize truth by intuition. It is certainly an adequate explanation, and, in the absence of a better one, we are logically driven to its provisional acceptance.

Here, then, we find another psychic phenomenon of the most stupendous proportions and of the most far-reaching significance; for it is participated in by all Christendom, and the subject-matter involves the most momentous problems of human life; indeed, it may be added that the ethical doctrines of Jesus are universally accepted wherever they are known, whether in Christian or in pagan lands. This part of his teachings may be summed up in these words: The universal brotherhood of man, charity for the poor and unfortunate, peace on earth, and love and good-will to all mankind. No one disputes the soundness of these principles, or doubts their universal practicability as a code of ethics for all humanity. Jesus was the first to teach them in their entirety. The Golden Rule, it is true, was formulated many years before the birth of Christ; but the idea of mankind as constituting one universal brotherhood, the children of one God, was his; and so was the doctrine of charity, peace, love, and good-will. It was these doctrines that first broke down the barrier between the Jews and the Gentiles and between the black and the white, and that has since struck off the shackles from untold millions of slaves, mitigated the cruelties of war, promoted the arts and sciences, justice and benevolence, freedom and good government, and established as the chief corner-stone of our civilization the idea of the sanctity of human life and the inalienability of human liberty.

What I have said of the ethical doctrines of Jesus applies with almost equal force to his whole system of religion. His fundamental idea of the Fatherhood of God, and his doctrine of the immortality of the soul, when added to the ethical principles before mentioned, may be said to constitute the essential features of his whole system of ethics, morals and religion. And it will not be denied that, as a whole, they appeal strongly to the unperverted intuitions of all mankind. Indeed, there is practically but one of his doctrines that has ever been seriously disputed; namely, that of the immortality of the soul. No one disputes the existence of a higher power to which all things are subject. The differences of opinion concerning that power are merely different conceptions of its attributes. Pantheism is but a variety of theism, and atheism really exists only in name. Science has disputed the doctrine of immortal life largely because it has been asked to accept it on faith alone; that is to say, because the proofs offered have been inadequate from the standpoint of material science. It is, nevertheless, true that the human soul instinctively recognizes the truth of every essential doctrine that Jesus promulgated.

I have spoken in previous chapters of the "evolution of the spiritual man." It would have been equally appropriate to designate the various epochs I have mentioned, as steps in the evolution of religion; for they are but different aspects of the same subject-matter. Considered as steps in the evolution of spiritual humanity, the process still goes on, and must go on until perfection is reached, until all humanity reaches the altitude of spiritual development attained by Jesus himself. Indeed, the evolution of the spiritual man is, in one sense, but a step in the great process of organic evolution. It is the final step in that process of development which began in protoplasm and culminated in man. I say "culminated in man;" for the same process of reasoning, the same series of phenomena, which demonstrates the scientific truth of the doctrine of organic evolution, proclaims man as the highest creature that can ever have an existence on this earth, — as the goal towards which Nature tended from the beginning. Having attained that altitude, the process of zoölogical change came to an end, and henceforth the dominant aspect of evolution is, and must henceforth be, in the direction of intellectual and spiritual progress and development.

Considered as steps in the evolution of religion, the same series of phenomena which we have been considering culminated in the religion which Jesus taught. And that was the end of what may be termed the organic evolution of religion. It reached its highest possible altitude in the simple but grand and all-comprehensive code embraced in

Christianity. By the term "Christianity" I do not mean that vast mass of theological doctrine evolved by Augustine, Athanasius, Clement, Justin Martyr, and Tertullian; nor do I refer in the remotest degree to that mass of dogma so ingeniously aggregated by the lesser lights of later years, which has usurped the title of Christianity. I mean the pure and simple code of morals, ethics, and religion — the real and essential Christianity — which fell from the lips of the man of Nazareth. I repeat, that was the end of the evolution of religion on this earth; for in that code perfection was attained. No one has ever succeeded in improving upon it. No one has ever been able to conceive a higher standard. We hear much of "the religion of humanity" from those who would free themselves from the restraints of the creeds and dogmas of the Church; but the "religion of humanity" owes its principles to Jesus, and to him alone; and the highest ideals of altruism find their realization in the same perfect character. Says Renan:

> "Jesus founded the absolute religion, excluding nothing, determining nothing, save its essence. . . . The foundation of the true religion is indeed his work. After him there is nothing more but to develop and fructify." [1]

The only attempt that has ever been made to find a vulnerable point in the doctrines of Jesus has been in the form of a declaration that the ethics of the Sermon on the Mount "are too good for this world." It may be true that some of his precepts are impracticable in the present state of civilization. It may be that the meek shall not inherit the earth for many long years to come. But the process of the evolution of humanity towards a higher civilization has not yet ceased; and we may rest assured that the time is approaching when there will be universal "peace on earth and good-will to all mankind." The religion of Jesus is for all time to come. It is the religion of the poor and the

[1] Life of Jesus.

lowly, and it is adapted to the highest civilization conceivable by man. It is the final religion of humanity; and though the earth in the fulness of time may pass away, his words shall not pass away. This is why I have remarked that the evolution of religion ceased when Jesus promulgated his doctrines. It had attained perfection; and that is all that evolution can do. It is true that his teachings have been misunderstood and perverted, and for many long years the evolution of religion has progressed backward. A vast system of theology has been erected, ostensibly upon the foundation which he laid, — a theology much of which bears no resemblance to true Christianity. But this was because man was, as he still is, imperfect. As civilization progresses, however, man will be released from the thraldom of creed and dogma, and revert to the pure and simple code of the man of Nazareth. "For other foundation can no man lay than that is laid." [1] "After him there is nothing more but to develop and fructify."

As in the organic world the highest possible type is man, so in the religious world the highest possible type is Christianity; and all future evolution of man or of religion must be in the direction of a higher civilization, — a more perfect manhood, with all that the name implies.

[1] 1 Corinthians iii. 11.

CHAPTER X.

PSYCHIC PHENOMENA OF PRIMITIVE CHRISTIANS.

Spiritistic Phenomena among the Early Christians. — Testimony of the Christian Fathers. — The Departure from Jesus' Example. — Paul's Explanation of Spiritistic Phenomena. — John's Tests. — Paul's Ecstatic. — The Oriental Ecstatics. — Modern Occidental Ecstatics. — Alleged Perception of Divine Truth in the Ecstatic Condition. — Neither Jesus, Paul, nor John believed in Spiritism. — Primitive Christianity promoted by Psychic Phenomena. — Constantine. — The Priesthood. — Prohibition of Psychic Manifestations among the Laity. — The Beneficence of the Inhibition.

IT would be interesting and perhaps profitable to trace the history of psychic phenomena from the time of Jesus down through the dark ages, and to note its influence upon the Christian Church both before and after the days of Constantine. But, fortunately for the common people, the production of the phenomena, after the first three hundred years of primitive Christianity, was confined largely to the priesthood, — that body having set up a claim to the exclusive right to work miracles, by virtue of their claim to the apostolic succession. The result of this was that its production was diverted to vastly different uses from those contemplated by the Master, and its history is, consequently, so contorted and obscured that it would be difficult to separate the genuine from the spurious.

It will be comparatively easy, however, to discover the influence which psychic manifestations exerted upon the early Christians, and to speculate with some degree of accuracy upon the effect which phenomena cognate to if not identical with those of modern spiritism had upon the destinies of the Church and the character of its teachings; but, in a work like the present, even this can be but briefly alluded to.

It is well known that such phenomena began to be produced among the early Christians almost immediately after the Crucifixion, and continued to be a salient feature of Church customs, certainly until the days of Constantine. This fact is abundantly attested by the writings of the early Christian Fathers, healing of the sick by the laying on of hands being one of the most common of the manifestations of psychic power. This power was not then confined to any one class or rank, but was possessed by all who observed the conditions prescribed by the Master. The physical condition necessary for the most successful work of this kind being identical with that required for the production of other phenomena, it soon became a common practice to go through with the whole repertoire of what are now known as spiritistic phenomena. Saint Paul himself mentions a long list of such phenomena which were produced in his day; [1] and Ignatius has this to say, —

> "Some in the Church most certainly have a knowledge of things to come. Some have visions, others utter prophecies, and heal the sick by laying on of hands; and others still speak in many tongues, bringing to light the secret things of men [telepathy] and expounding the mysteries of God."

Saint Anthony declared that, after fasting, he had often been surrounded by bands of angels, "and joyfully joined in singing with them." Tatian declares that "our virgins at the distaff utter divine oracles, see visions, and sing the holy words that are given them," being "full of the faith in Christ." Tertullian relates the case of a sister in the

[1] See 1 Corinthians xii.

Church, who, when entranced, was able to see spirits; and Montannas affirms with great emphasis that prophecies, the power to heal the sick, "tongues and visions, are the divine inheritance of the true Christian." These statements are amply confirmed by Apollinaris, Barnabas, Clement, Cyprian, Lactantius, Papias, and others. It was a common event in these manifestations for their psychics to hold alleged communication with the angels; and Tertullian declares that, during religious services, they became entranced, and sometimes "beheld Jesus himself, heard the divine mysteries explained," and "read the hearts" of those present.

It is almost superfluous to observe that these manifestations were identical with the so-called spirit manifestations of the present day. But it is worth while to note the fact that not one of them was authorized or countenanced by Jesus, with the single exception of that of healing the sick. This is a most significant fact, and it is demonstrative evidence that he discountenanced the practice, knowing, as only he could know, that communication with spirits was impossible. He knew the laws governing all such manifestations, and it is to the last degree improbable that he would have neglected to instruct his followers in the art of spirit intercourse, if by that means they could have been put into communication with intelligences capable of "explaining the divine mysteries." It is also to the last degree improbable that one whose mission it was to "bring life and immortality to light" would have neglected so glorious an opportunity to demonstrate the truth of his teachings, and to point out a means by which his disciples could hold communion not only with angels and ministers of grace, but with himself after he had ascended to the Father. His whole life and career was a living protest against that species of psychism wherein the prophets assumed to have direct verbal communication with God, and others claimed to hold communion with spirits of the dead; the latter, however, being denounced as witchcraft by the Mosaic law and punished with death. Is it not probable and in keeping with the whole character and the career — the mission — of Christ, which was to teach spiritual truth to mankind, that, if communication with spirits of the dead had been possible, and if it had been that beneficent practice which modern spiritists would have us believe it to be, he would have in some way indicated to us his approval of such practices? If it is true that spirits of the dead can communicate with the living inhabitants of this world, he knew it. If it is true, it is important for us to know it; for that would be demonstrative of a future life. If it is demonstrative of a future life, he would surely have informed us of the fact, and would have enjoined upon mankind a diligent cultivation of the art of spirit intercourse. It was his mission to teach the doctrine of immortality. It was his desire and purpose to demonstrate the fact of immortality; and he accomplished his object so far as it was possible for him to do so in the age in which he lived. He has left a record which gives us indubitable evidence of his perfect knowledge of the laws of the human soul. He has left a record demonstrative of his perfect character and of his zeal for the promulgation of spiritual truth. He offered up his life as a sacrifice upon the altar of spiritual truth. The spirit of altruism was regnant in his whole character; but if there was one thing more than another wherein that spirit was manifest, it was in his desire to teach to mankind the fact of immortality. It is simply a monstrous absurdity to suppose that, if it was possible to communicate with departed souls, he deliberately neglected so grand an opportunity to demonstrate the truth of the essential doctrine which it was his mission to bring to light; and that it was left for hysterical women of the nineteenth century, aided and abetted by convulsive furniture, to teach us "the way, the truth, and the life."

It is impossible to suppose that he was not aware of the psychic manifestations of his day, and of the current theory of their origin.

"The group," says Renan, "that pressed around him upon the banks of the Lake of Tiberias believed in spectres and spirits. Great spiritual manifestations were frequent. All believed themselves to be inspired in different ways."

But there is no record to show that he did more than to tolerate the current beliefs. He did not sanction them either by precept or by example; much less did he encourage them by advancing the idea that the phenomena proceeded from disembodied spirits. On the contrary, his whole life was a protest against such beliefs and such practices. By precept and example he taught the world that healing the sick was the only legitimate use of psychic power; and the lesson of his three temptations in the wilderness is that neither for bread, nor for glory, nor for power, nor for emolument, can psychic power be legitimately exercised outside of the limitations which he prescribed.

After the crucifixion and death of Jesus, Saint Paul appears to have been tolerant of the psychic manifestations which soon became common in the Church, doubtless for the reason that it was, ir that primitive and superstitious age, an element of strength. It enabled Christianity to become an aggressive power, carrying with it what was supposed to be demonstrative proofs of its divine source in the form of phenomena the supermundane origin of which in that day could not be successfully denied. To the credit of Saint Paul, however, it must be remarked that he not only had a very clear perception of the true origin of the phenomena, but he took pains to place on record a statement of his convictions. Paul was a learned man, filled to saturation with the philosophies of the civilized world; and although he sometimes injected some of his Greek philosophy into that of Jesus, yet he was a man who could not be deceived as to the true origin of the spiritistic manifestations; and he took particular pains, in his first epistle to the Corinthians, to disabuse their minds of the idea that the phenomena which at the time appear to have constituted a salient feature of Christian worship, had their origin in spirits of the dead. In the first eleven verses of the twelfth chapter of first Corinthians he discourses as follows: —

> "Now concerning spiritual gifts, brethren, I would not have you ignorant.
>
> "Ye know that ye were Gentiles, carried away unto these dumb idols, even as ye were led.
>
> "Wherefore I give you to understand, that no man speaking by the Spirit of God calleth Jesus accursed: and that no man can say that Jesus is the Lord, but by the Holy Ghost.
>
> "Now there are diversities of gifts, but the same Spirit.
>
> "And there are differences of administrations, but the same Lord.
>
> "And there are diversities of operations, but it is the same God which worketh all in all.
>
> "But the manifestation of the Spirit is given to every man to profit withal.
>
> "For to one is given by the Spirit the word of wisdom; to another, the word of knowledge by the same Spirit;
>
> "To another, faith by the same Spirit; to another, the gifts of healing by the same Spirit;
>
> "To another, the working of miracles; to another, prophecy; to another, discerning of spirits; to another, divers kinds of tongues; to another, the interpretation of tongues;
>
> "But all these worketh that one and the selfsame Spirit, dividing to every man severally as he will."

It would thus appear that Paul had formulated a working hypothesis regarding all spiritistic phenomena, the essential features of which were, first, the repudiation of the prevalent idea that the different manifestations of spirit control arose from communion with a corresponding number of disembodied spirits; second, the broad assertion that all such phenomena proceeded from the same source, namely, the spirit of God manifest in and through that part of Him which constitutes the soul of the psychic who produced the phenomena. There is, however, one phrase in the foregoing quotation which requires a word to make it clear; namely, "to another, the discerning of spirits." This has been held to imply an acknowledgment by Paul of the existence of the power to see spirits. This confusion arises from a mistranslation of the text. Instead of "discerning," which implies the exercise of the physical function of seeing, the word should be *discrimination,*[1] which implies merely the mental faculty of sound judgment. With this view of the case, it would seem that the power of discrimination, when applied to the divers gifts and manifestations mentioned in the context, was the most desirable of the whole repertoire; especially when we take into consideration "the diversities of gifts," "the differences of administrations," and the "diversities of operations," together with the law of suggestion, which was just as potent a factor in their psychic phenomena as it is in ours. John, however, greatly simplified the process of "discrimination" in such matters. The following comprises his formula: —

"Beloved, believe not every spirit, but try the spirits whether they are of God; because many false prophets are gone out into the world.

"Hereby know ye the Spirit of God: Every spirit that confesseth that Jesus Christ is come in the flesh, is of God:

"And every spirit that confesseth not that Jesus Christ is come in the flesh, is not of God. And this is that spirit of antichrist, whereof ye have heard that it should come; and even now already is it in the world."

Thus it will be seen that John was in perfect accord with Paul in his method of "discrimination of spirits." In this connection it may be well to remark that the phrase "try the spirits" has often been held to imply that John was a believer in spiritism. But it is obvious that he employed

[1] See Rotherham's Literal Translation.

the word in a sense that does not warrant such an inference. He mentions but two spirits; namely, the spirit that is of God, and the "spirit of antichrist." This clearly indicates that he employed the term to designate a mental condition or disposition, — an intellectual or moral state, and not a disembodied soul. Indeed, the one definition necessarily excludes the other. Moreover, on general principles it may be definitely affirmed that John was not a believer in spirit communications from the other world; for he was a disciple of Jesus, and had imbibed instruction from the fountain-head. Spiritists, both ancient and modern, are fully persuaded that they are in complete possession of accurate knowledge of the world to come and of the general internal economy of Heaven. On the other hand, Jesus did not pretend to know; or if he did, he consistently refrained from imparting that information to his followers, except in the most general terms, which will be noted hereinafter. Hence it was that John, notwithstanding his intimacy with the Master, was forced to confess that he knew nothing of what is in store for us on the other side. "It doth not yet appear," said he, "what we shall be; but we know that, when he shall appear, we shall be like him; for we shall see him as he is." [1]

Saint Paul has also been accused of spiritism, the accusation being based upon the passage wherein he says, —

> "I knew a man in Christ above fourteen years ago, (whether in the body, I cannot tell; or whether out of the body, I cannot tell [i. e. whether his soul left the body or not]: God knoweth;) such an one was caught up to the third heaven.
>
> "And I knew such a man, (whether in the body, or out of the body, I cannot tell: God knoweth;)
>
> "How that he was caught up into paradise, and heard unspeakable words, which it is not lawful for a man to utter." [2]

Those who are acquainted with the literature of *ecstasis* will readily understand that Paul was describing a person

[1] 1 John iii. 2. [2] 2 Corinthians xii. 2-4.

who was in that psychic condition known as "ecstasy." It is a state of profound hypnosis, the deepest that can be produced with safety to the subject, and the consequent phenomena depends, as in all other grades of hypnotism, upon the dominant suggestion in the mind of the subject when he enters the state. It is frequently self-induced, especially among the East Indian adepts and Yogis, many of whom spend the greater part of their lives in that condition, or for preparing themselves for entering it. They induce it by sitting in one attitude for an indefinite length of time and thinking about themselves; the latter part of the process being by them denominated "introspection." By this means a state of profound hypnotism is induced, accompanied by an equally profound and all-comprehensive egotism, the result aimed at being what is known in their vocabulary as "illumination." This state of "illumination" appears to be the culmination of the ecstatic condition; that is, they have reached a point where egotism can go no farther. It is then that they identify themselves with the forces of Nature, and imagine that they can thunder and produce earthquakes and other cataclysms, and that they are in possession of all knowledge and power and all dominion.

It may be remarked, in passing, that upon this phenomenon is based the Oriental claim to superior knowledge of science and of the laws of Nature which is so confidently set up by those self-immolated victims of subjective hallucination. It is this mental condition, entered into in utter and profound ignorance of the fundamental law which governs all psychic phenomena, that gives rise to their lofty contempt for Western science and civilization. They never tell us what they have seen or what spiritual secrets they have penetrated in this state of "illumination," but declare that "the world is not ready to receive it," etc., etc., and bid us wait until they decide that the time is auspicious for them to yield up the sum total of the secrets of Nature. About the cosmogony of the physical universe, however, they are not always so reticent, but will cheerfully inform us that out of their "cosmic consciousness" they have evolved many truths of the utmost importance, such as that the earth is the centre of the physical universe, and that the sun revolves around it; that there is to the north of us a great mountain behind which the sun, moon, and the other planets retire in their turn to rest; that eclipses of the moon are caused by "dark planets" coming between the earth and the moon,[1] and many other such "truths," of which Western science has been for many years profoundly ignorant.

But this is a digression. The phenomenon has often been reproduced even in the unilluminated Occident. It is often caused by nervous prostration, and it can be produced by hypnotism, or by hasheesh, or by any of the processes usually employed in the induction of the subjective condition; and, as in all other psychic states, the visions beheld, or the impressions experienced, are in exact accordance with the suggestions which dominate the mind of the subject when he enters the state. If there is no specific suggestion made, the topic upon which the subject will be "illuminated" will be determined by the dominant characteristics of his mind. Thus, an inventor will feel that he has attained the power of perception of all mechanical laws and forces, and that all problems of invention are as a b c to him; the mathematician will feel that he is in possession of all the laws of numbers, and that any and all problems are easy of solution by his "illuminated" intelligence; the musician will experience the feeling that he is in an atmosphere of musical sounds, and that the most delightful harmonies await his volition; and so on through the whole repertoire of human accomplishments and objects of earthly ambition. Again, if the suggestion is made that the ecstatic

[1] See Carpenter's "From Adam's Peak to Elephanta," p. 186.

shall visit the spirit land, — the abode of the blest or the dungeon of the damned, — he will, with equal facility, visit either place, and his account of his visions will be the exact reproduction of his preconceived opinions (auto-suggestions) on those subjects.

But here is the peculiarity which attends the visions and impressions of a certain class of ecstatics the world over. When they enter that state with the dominant idea that they are going to come into contact with Omniscience and merge their intelligence, as it were, with that of the Deity, they have no specific idea of what they are about to see, to experience, or to learn. Their only suggestion is a general one to the effect that they are about to come into intimate and loving contact with the All-Knowing One; and that, guided by the very Spirit of Truth, they will be able to obtain instant possession of all knowledge. The result is always the same, from the Oriental ecstatic down to the humblest hypnotic subject. They experience a sensation which is described to be to the last degree pleasurable and exhilarating, — a feeling that they have been suddenly released from the trammels of the flesh, that they are "emancipated" and "illuminated," and that they are in possession of all truth and all power and dominion. But that is the sum total of their revelations. No one has ever been able to wring from them the smallest modicum of that vast store of cosmic intelligence of which they have so suddenly become the custodians. One amiable old gentleman, known to the writer, has become so full of enthusiasm over his own experience in the ecstatic condition that, realizing the hopelessness of any attempt to convey by words the remotest idea of the "great truths" of which he has become the depositary, he has started in to reform the world by inducing others to enter the ecstatic condition, so that each for himself may acquire possession of that vast fund of unspeakable information which belongs alone to the "illuminated."

"It cannot be conveyed in words," is the reply of the average Occidental ecstatic when pressed for an account of his experience. The Oriental wraps himself in his mantle of mysticism, and refuses to reveal the awful mysteries which have been confided to his care; but boasts of his vast storehouse of "scientific" knowledge, and looks with contempt upon the frivolities of material science in the Occident. Saint Paul's ecstatic evidently had the experience and was hedged about by the same limitations. During his visit to paradise he "heard unspeakable words which it is not lawful for a man to utter," says Paul; and it must be admitted by the most sceptical that this was a double restriction not easy to evade or overcome.

Well may Paul have been in doubt as to whether the man "was in the body or out of the body" when he was in that profound trance condition which often simulates death, and which is the condition necessary for the production of the phenomena of ecstasis. His doubt must have referred to the man's physical condition, and not to any question of the corporeal existence of the man himself; for Paul explicitly states that he knew the man.

I have discussed this matter at greater length, perhaps, than was necessary; but this passage has been so often tortured into the service of spiritism that I have deemed it expedient to classify the fact in accordance with the views of modern experimental psychology, and at the same time to relieve Paul from the charge of inconsistency, in view of his explicit declaration that the alleged spiritistic phenomena than prevalent in the Church were neither more nor less then manifestations of one and the same Spirit; namely, the spirit of God.

Besides, I deemed it important to show that Paul, the most learned and philosophic of the Apostles, was not tinctured with the then prevalent beliefs; and that John, the most intimate personal friend and companion of Jesus, was also free from the prevailing delusion. The importance of these facts is seen from the inference, which is irresistible, that, if Jesus had deemed it possible for disembodied spirits to communicate with the living, his views would have been reflected in the teachings of those Apostles. The fact that neither Jesus himself nor John, his friend, nor Paul, the most learned exponent of his philosophy, nor indeed any of his Apostles, ever intimated a belief in spiritism, is conclusive against the hypothesis that the Founder of Christianity regarded spirit intercourse with the living as a possible factor in the science of the soul.

Nevertheless the fact remains that the phenomena of so-called spiritism constituted one of the salient features of primitive Christianity. And it is no discredit to the Christian religion to say that these phenomena constituted one of the most potent agencies employed for its promulgation. Indeed, it may be doubted whether the Christian Church could have long survived as an organic institution in that day and age of the world, had it not been for the signs and wonders which were afforded by the various forms of psychic phenomena which were then prevalent. Jesus himself recognized the necessity for thus satisfying the popular demand for evidences of his authority and his divine mission. It is true that he produced none outside of a clearly defined limit; and that limit was defined by the spirit of altruism regnant in his soul. His miracles were all wrought for the benefit of humanity, — for human enlightenment or for the relief of human suffering. Their effect as evidences of his divine mission was of secondary importance to him. He did nothing for display, nothing for glory, nothing for emolument, nothing even to convince the sceptical of the truth of his doctrine, unless he could at the same time confer a benefit upon suffering humanity. He enjoined upon his followers the duty to heal the sick, and left his example as a sacred heritage to all who should come after him.

During the first few hundred years of the Christian dispensation his injunctions were faithfully observed. But the power to heal the sick implied the power to produce other phenomena; that is to say, the psychical condition necessary for the production of other phenomena were necessarily induced by the training required for the acquisition of the power to heal the sick. The result was that other phenomena were produced. It is superfluous to say that the same laws that prevail to-day governed the production of psychic phenomena in that day. The law of suggestion exerted its subtle influence then as now. Consequently the same facilities for self-deception on the part of the psychic existed in the Church of that day as exist in the spiritistic circles of the nineteenth century. The conditions were the same. The phenomena were identical. The same tests were applied with the same wonderful results. The phenomena, under the subtle influence of the law of suggestion, lent itself to the confirmation of every belief, just as it does in the spiritistic séances to-day. The necessary result was that the psychics of the Church, being dominated by the suggestions embraced in the Christian faith, confirmed the beliefs of the Church.

It is easy to see what a powerful proselyting engine was at the command of the primitive Church, and to account for the zeal and success of that simple-minded people. Says Mosheim, —

> "It is easier to conceive than to express how much the miraculous powers and the extraordinary divine gifts which the early Christians exercised on various occasions, contributed to extend the limits of the Church."

Possessed of such "divine gifts" emanating from a source which they could but regard as supernatural, in constant communication with beings whom they could but regard as disembodied spirits, frequently beholding visions of a sublime figure which they could but believe to be that of the

Divine Master himself, every phenomenon confirming and emphasizing the central idea of their faith, it would have been a miracle indeed if the Church had not flourished and become a moral force in the world, potent, aggressive, irresistible. And such we find it to have been; for their missionaries, in spite of pagan persecution, were soon thundering at the gates of Rome, Alexandria, and Constantinople. The result was that in the short space of three hundred years they had become so powerful and influential that Constantine thought it worth while to become converted to Christianity, to the end that he might establish himself as Emperor and convert Christianity into a state religion.

With that event ended the golden days of primitive Christianity. Evolution had turned her face to the rear. The pure and simple doctrines of Jesus were forgotten. His text of discipleship — "that ye love one another" — was discarded. Thenceforth Christianity was largely a thing of creed and dogma. The Church was given over to the control of an organized priesthood who acknowledged allegiance only to the State. Private opinion became a public crime. The reign of love which Jesus inaugurated was transformed into a reign of terror; and for more than a thousand years the pathway of the Church was illuminated by the fagot and defined by human blood.

Obviously it would be difficult, if not impossible, to trace the true history of psychic phenomena through that period. The heritage which Jesus had bequeathed to all his disciples had been seized by the priesthood and made to subserve its interests and to promote its power. That history, therefore, was in the keeping of those whose interest it was to deepen its mysteries, to the end that the people might be kept in ignorance of its source. That the priesthood have been, from time immemorial, in possession of the power to produce most wonderful psychic phenomena, is well known in the inner circles of the Church. That they are not entirely ignorant of its true source, is apparent. It is a part of esoteric Romanism.

These remarks are not made in any spirit of censure; for if the Church has ever done one thing more praiseworthy than another, it was when it inhibited the production of spiritistic phenomena by the common people. No matter what secret motives may have actuated the priesthood in confining the production of psychic phenomena to that order, the fact remains that if the common people had not been prohibited from the indiscriminate production of psychic phenomena, it would have utterly demoralized the Christian Church, and rendered it a very cesspool of vice and immorality. No one who has investigated the subject needs to be told how demoralizing to soul and body is the production of spiritistic phenomena, even in this enlightened age, especially where the medium is ignorant of its true source, and ascribes it to supermundane agency. How much more terrible would have been the results in an age of universal ignorance and superstition, can only be conjectured. In the early days of the Christian Church learning was confined largely to the priesthood; and it is doubtless true that they early discovered the vicious tendency of such practices, and felt compelled to interfere, in the interest of morality, and to prohibit the indiscriminate production of psychic phenomena by the ignorant laity. It is also doubtless true, as before remarked, that the priesthood understood something of the true nature of psychic phenomena; and that they should employ it occasionally for the promotion of the interests of their order, was inevitable. The people of their day were seeking for "signs and wonders" as they were in the days of Jesus, and the priesthood had before them the example of the Master in withholding from the laity the esoteric knowledge which they were prepared neither to receive nor to appreciate.

Thus it appears manifest, not only that it was through psychic phenomena that the Christian religion was evolved, but that it was largely through psychic phenomena that the Christian Church was enabled to establish itself on a firm basis within the three hundred years succeeding the crucifixion of its founder. It matters not that these psychic phenomena were misunderstood; nor is it any discredit to Christianity that in an age of intellectual darkness, before science threw its first glimmering rays of light upon the intellectual horizon, the Christian religion was thus promoted. Jesus did not misunderstand the phenomena of the soul, nor is he to blame because his followers mistook the import of phenomena which he did not produce and to which he did not give his sanction. As well might we discredit astronomy because it had its origin in astrology, or chemistry because it was preceded by alchemy, or impugn the wisdom of the Almighty because the Psalmist was ignorant of the Copernican system when he exclaimed, "The heavens declare the glory of God, and the firmament showeth his handiwork." The grand procession of the planets around the sun constitutes the phenomena of the solar system. These phenomena are the facts of astronomy; and they are none the less so because they have been misunderstood. They were observed and studied alike by the ignorant and the wise until the truth was evolved. The Psalmist doubtless regarded the earth as the centre of the universe; but the discoveries of Kepler and of Newton have neither diminished the force nor discredited the truth of the sublime words of the sweet singer of Israel.

In like manner psychic phenomena — the facts of the science of the soul — have been observed from time immemorial, and must still continue to be observed and studied until the true science of the soul is evolved. When this is accomplished, it will be found that the truths of

Christianity will be none the less clearly recognizable because psychic phenomena have been, in times past, most grossly misinterpreted. On the contrary, as every truth illuminates every other truth to which it is related, the truths which Jesus taught must find a new illustration in every fresh discovery in the science of the soul.

CHAPTER XI.

MODERN PSYCHIC PHENOMENA.

Mesmerism. — Telepathy demonstrated by the Followers of Mesmer. — Braid's Discovery. — Hypnotism. — Discovery of the Law of Suggestion. — Clairvoyance. — The Rochester Knockings. — Mesmeric Subjects and Mediums. — Spiritism as a Step in the Process of Evolution. — Its Effect.

WE now approach an epoch in the history of psychic phenomena of the most transcendent interest and imminent importance. Hitherto we have dealt with phenomena so obscured by the twilight of tradition and imperfect history that only the faint outlines or the most salient features have been discernible. We shall, however, be compensated for this lack of clearness in the ancient phenomena by turning upon it the calcium light of modern experience; for we now enter the domain of demonstrable facts which have been incorporated into the great body of modern science; namely, the scientifically verified facts of experimental psychology.

When Anton Mesmer first demonstrated to the world that by certain mysterious manipulations persons can be thrown into a condition of trance, during which the objective senses are held in more or less complete abeyance, and that at the same time the functions of the body can be modified, pain suppressed, fever calmed, and disease removed, he laid the foundation of the true science of the human soul. Not that he obtained more than a glimpse into the promised land, or that he had the remotest idea of the grand results which were to follow; but he was the first, in modern times, to point out one mean through which the soul can be experimentally studied. For him the only field of usefulness for the newly discovered power was that of therapeutics; and it was not till he had been driven by professional jealousy into dishonored exile that his followers so far extended his discoveries as to open the way for the study of the whole field of experimental psychology. The Marquis de Puységur, a philanthropist, a scientist, and a man of fearless integrity, in utter disregard of the sentence of professional and social ostracism pronounced by the medical profession of his day upon all who presumed to investigate the subject of mesmerism, extended the experiments of Mesmer, and was the first to develop experimentally the phenomenon of telepathy. He was followed by many others, of more or less scientific prominence, who confirmed his experiments, among whom were Esdaile, Elliotson, Deleuze, Baron Dupotet, and many others of lesser note. The result was that a series of most wonderful psychic phenomena were produced and verified with scientific exactitude. The ultra-scientists, however, continued to cast ridicule upon the phenomena, and to persecute and to drive to ruin every scientist who dared to make an honest experiment. This continued until 1840, when Dr. Braid, a Manchester physician, announced that he had discovered that a condition cognate to that produced by Mesmer could be induced by causing the subject to gaze steadily upon a bright object held in front of and slightly above the eyes. This he denominated "hypnotism;" and the name has since been retained and applied to all the varied phases of induced subjective phenomena, although it strictly applies to but a very small proportion of them. His work, however, attracted very little immediate attention in his own country, and it was not until Liébault confirmed and extended the experiments of Braid that hypnotism was admitted within the domain of the exact sciences. Liébault was followed by a host of others, so that within the last twenty years, or less, hypnotism has come to be acknowledged as a science of the most transcendant interest and importance, not alone in its aspects as a therapeutic agent, but as the handmaiden, *par excellence*, of experimental psychology.

It is to Liébault, however, that the world is indebted for the greatest discovery ever made in the science of hypnotism; namely, the law of "suggestion." This law, reduced to its simplest terms, is that "persons in an hypnotic condition are constantly amenable to control by suggestion." I am aware that the credit of this discovery has been claimed for Braid; but, nevertheless, Liébault first formulated the law, and it is to the one who has the capacity to grasp the universality of a law, and definitely to formulate it, that credit is due for the discovery. Many others had noted the effect of suggestion in particular cases, and had thus counted it as a possible factor in hypnotism. Paracelsus, in the sixteenth century, noted it as an important factor in psycho-therapeutics; but he was no more entitled to the credit of its discovery than were the predecessors of Newton, who talked learnedly of gravitation but died before its fundamental law was formulated, entitled to the credit due to the author of the "Principia." It is true that Liebault confined his formula to the phenomena of experimental hypnotism, and it was left for later investigators to discover that the law was the universal and dominating factor in all the multiform phases of psychic phenomena. But until Liébault's discovery was definitely formulated, hypnotism was not, nor could it be, entitled to admission into the circle of the exact sciences. That discovery, therefore, constituted the first great step in the evolution of the science of the soul; for it is by hypnotism and cognate phenomena alone that the fact that man has a soul can be scientifically demonstrated.

In the mean time, before the discoveries of Braid and of Liébault, mesmerism had fallen largely into the hands of ignorant charlatans, who travelled throughout every civilized country, lecturing and giving exhibitions of the phenomena to gaping crowds, who were neither more nor less capable than were the exhibitors themselves of appreciating the real significance of the exhibitions. By this means mesmeric subjects were indefinitely multiplied until every little hamlet in Christendom could boast of its "seers" and its "prophets." "Clairvoyance" became a word of portentous import, and was soon employed as a generic term for every manifestation of perception not directly traceable to sensorial experience. Many notable phenomena were produced, and books describing them multiplied. All were, of course, derided by the "scientists;" the lecturers themselves were branded as "mountebanks," and the "subjects," who were taken largely from the ranks of the school children, were denounced as "frauds," "humbugs," and "swindlers." Among the books which appeared, many were written in a purely scientific tone and spirit, and described the experiments with great minuteness and exactness, and detailed the tests applied and the safeguards employed with true scientific caution and transparent honesty of purpose. It is a curious and an intensely interesting study to go over those old "volumes of forgotten lore," rescued from the top shelves of second-hand bookstores, and to compare the experiments therein detailed with those of the scientists of to-day. Such a study reveals many thoroughly authenticated facts of great scientific value in the study of experimental psychology, the only defect being that the law of suggestion had not yet been formulated. Nevertheless, in very many cases, that factor was as intelligently eliminated as it has ever been by later scientists who are in full possession of a thorough knowledge of its potency and universality. In point of fact, there are few phenomena of importance produced by the later scientists which have not their counterpart in the old records of pre-spiritistic mesmerism.

In this connection it may be well to pause for the purpose of remarking that up to a certain date no one ever dreamed of ascribing to supermundane agency any of the phenomena of mesmerism. Many phenomena were produced which have since found a ready solution in the hypothesis of spirit intercourse; but at the time of their production it was not found necessary to presuppose any other agency than that of some inherent, though newly discovered, power of the living subject. In other words, it was not regarded as necessary for a man to be dead before he could develop the powers evoked by mesmerism. Such an hypothesis had not yet been "suggested."

This state of affairs, however, was destined to a somewhat sudden termination. The Rochester knockings had commenced. At first ascribed to trickery, investigation proved that they could not be traced to any known physical agency. The sounds were startling, mysterious, uncanny, and to none more so than to those to whose unconscious agency they were afterwards traced. To the minds of the local savants the most obvious solution was the supernatural. This idea once suggested, a test was easy. A code of signals was improvised, and the raps were questioned. An intelligence was found to be behind the mysterious sounds. On cross-examination, that intelligence freely admitted itself to be none other than that of a disembodied spirit; and the raps first made upon the walls of the humble residence of the Fox sisters were heard around the world.

A new era in psychic phenomena had been inaugurated. In an incredibly short space of time "mediums" of communication with the inhabitants of another sphere were found all over the civilized world. Mesmerism was forgotten. But it is a significant fact that it had unwittingly provided an abundant supply of the raw material for "mediums;" for every successful mesmeric subject was found to be already developed for successful mediumship. The phenomena of "clairvoyance" no longer possessed their former significance. What was, under mesmerism, the development of an inherent power of the mind of the subject, under spiritism was a message from some denizen of another world. If any doubt existed upon that point, it was speedily set at rest by the simple process of questioning the intelligence itself. When asked if it was a spirit, the answer was, "Yes." When asked if it was the spirit of John Smith, the answer was "Yes;" and the same answer would be returned if the identity of the spirit of Socrates was sought. In other words, it was just as easy successfully to invoke the shade of Socrates as it was to call up John Smith, notwithstanding the disparity in numbers.

The law of suggestion had not been discovered; and the fact of duality of consciousness existed in the popular mind only as a Platonic reminiscence.

But this is not the proper place to discuss the errors of spiritism. It is sufficient for present purposes to note that the appearance of spiritism, coming as it did upon the world of human thought and experience simultaneously with mesmerism, seems not only opportune, but almost providential. Together they constitute the great body of the psychic phenomena of the nineteenth century, and neither would have been complete without the other. In spiritism we have a vast series of phenomena, and in mesmerism and hypnotism we have a means of scientifically studying it and thus profiting by the lessons which it teaches. If this is done in a calm and dispassionate spirit, we may rest assured that what we shall learn will be for the highest good of the human race. We shall at least find that when we look upon it as a necessary part of the grand system of evolution of the human mind, it is a factor of inestimable value and significance. Viewed as a factor in the evolution of the spiritual man, it has been of transcendent value to mankind. It matters not that its phenomena have been grossly misinterpreted. It was impossible to avoid a misunderstanding of it in the absence of the knowledge which modern scientific investigation has revealed within the last decade. It would have been a miracle if it had not been accepted for all that it purported to be, in the absence of any other rational explanation than that afforded by a wholesale denial of its phenomena. Its adherents had daily ocular demonstration of the genuineness of its phenomena, which could not be offset by *a priori* denials from those who refused to investigate. Moreover, the intelligence behind the manifestations claimed to be that of loved ones who had gone before; and in the then state of human knowledge there was no means of successfully disproving the statement. Besides, it was a statement that millions of stricken hearts dreaded to have disproved. To many it constituted the last ray of hope of a life beyond the grave, and of a reunion with the loved and lost.

It does not always follow that the mistakes of humanity are productive of unmitigated evil. We have already seen how, in times past, the most grossly misinterpreted psychic phenomena led, by the slow but sure steps of evolution, to a knowledge of the true God; and how the propagation of the true religion was promoted by the same means. Spiritism has also served a noble purpose in that it has stayed the wave of materialism which swept like a cyclone over the civilized world upon the announcement of the doctrine of organic evolution. Millions of the human family who could not appreciate the fact that the doctrine of evolution does not touch the question of true religion and leaves the problem of immortality just where it found it, have derived consolation from what they regard as demonstrative evidence of a life beyond the grave.

In making the foregoing remarks, I have not taken into

account any of the vagaries of spiritism. I have not considered the evils to which it has given rise, the gross immoralities which some of its votaries teach both by precept and example, nor the absurdities into which many of its followers have been led. When a law of Nature is misunderstood, there is inevitable danger to those who rashly place themselves within its reach. This is just as true of the laws of mind as it is of the laws of matter. The law which, rightly understood, is the most beneficent, may become an engine of destruction to those who ignorantly place themselves in wrong relations to it.

What I have here said of spiritism, therefore, must be considered as having reference to its aggregate effect upon the human family. In this respect I have no hesitation in saying that, as a whole, it has been beneficent. But, taking a still broader view of the subject, it must be said that its manifestations are a necessary and an indispensable part of the grand aggregate of psychic phenomena, through which alone man is at last enabled to study the science of the soul. Not that it teaches us just what is in store for man in a future state of existence, for, in the language of the Beloved Disciple, "it doth not yet appear what we shall be;" but it teaches us what man *is*. Not that we shall ever be able to enter into communication with the inhabitants of the spirit world, and thus learn more than Jesus revealed to us; but we may learn by induction something of what he knew by intuition; and we may, perchance, learn enough of the laws of the soul to be able to postulate immortality with some degree of scientific certainty. If this should prove to be the outcome of spiritistic phenomena, all will agree that it has not been produced in vain; even though, by the same process of reasoning, it should be demonstrated that spirits of the dead do not communicate with the living, and although the whole superstructure of spiritistic philosophy, based upon the assumption of spirit communion, should be demolished.

It seems to me that I am warranted in saying that enough of thoroughly verified facts have already accumulated to enable us to successfully apply the processes of induction to the solution of the problem of a future life. The facts of mesmerism; the facts of hypnotism, as developed by the scientific investigators of Europe and America; the vast array of scientifically verified facts presented in the reports of the London Society for Psychical Research, together with the rich store of facts presented in the phenomena of spiritism, — constitute the material from which it is hoped to learn something, not only of what man is, but of the fate to which he is destined.

CHAPTER XII.

HAS MAN A SOUL?

Intuitive Perceptions of the Existence of a Soul in Man. — Plato's Philosophy. — The Doctrine of Body, Soul, and Spirit. — The Doctrine of Jesus. — Modern Scientific Scepticism. — Requirements of Modern Science. — The Dual Hypothesis. — The Phenomena of Dreams. — The Objective and Subjective Mental States differentiated. — Limitations of Powers of Reasoning in the Subjective Mind. — Its Perfect Power of Deduction. — Telepathy and Prevision.

IT has thus far been provisionally assumed that man has a soul. But, before proceeding to formulate a scientific argument demonstrative of the soul's immortality, it is logically necessary to demonstrate the verity of the provisional assumption. Materialistic science will certainly be satisfied with nothing less; for it is at this point that it invariably calls a halt, and reminds us that the primary rule of logic demands that our premises be not assumed.

In discussing this branch of the subject in a work like this, the reader must be presumed to be somewhat familiar with the current literature relating to the psychic phenomena of the nineteenth century, especially with that which deals with the scientific aspects of the various questions involved. It is obviously impossible, within the limits of a single volume, to present documentary evidence of the verity of every statement that must be made. Therefore results only can be stated; but the reader may rest assured that I shall not attempt to lead him outside the realm of scientifically verified facts, which can be experimentally reproduced. Should I attempt to do so, the fraud would be easily detected, for the facts are public property, and are known to him who reads. My conclusions from those facts, however, rest upon a different footing. They are necessarily my own, and it is the province of the reader to test their soundness in the crucible of his own logic, if mine is found to be unsound or unsatisfactory.

That man has a soul, is and has been, since the dawn of civilization, a matter of intuitive perception; that is to say, all civilized people have felt that they realized, in a more or less definite way, that there is in man what appears to be a distinct entity which is apparently capable of sustaining an existence independently of the body. This fact of universality of perception constitutes a strong argument, though not a conclusive one, in support of that doctrine, and its corollary, — a future life. It is noteworthy that in the early history of the world, the higher the state of civilization the more pronounced and definite were the current notions regarding the soul's existence; although they lacked that clearness and simplicity which characterized the purely intuitional perceptions of those who were less skilled in philosophical ratiocination. Thus, in Greece, the doctrine was formulated in clear and definite terms by Plato who held that man is composed of "body, soul, and spirit." His doctrine, however, was the result of something more than intuition. He does not tell us the distinction between soul and spirit, and leaves us entirely in the dark as to whether his conclusions were arrived at from the observation of phenomena, or from purely speculative philosophy, without facts to sustain it. But it must be remembered that Plato, in common with the other philosophers of his day, regarded the conclusions derived from purely speculative philosophy as good as so many facts for the purpose of constructing the major premise of a syllogism. Not that they disdained facts, or failed to employ them when they were easily obtainable, but that they failed to estimate the relative value of a demonstrated fact and the conclusions resulting from their own speculations. Again, Plato may have been influenced by the Hindu philosophy, which constructs man in sections, puts him together like a telescope, and assigns him the task of shedding one section at a time until there is nothing left but "pure spirit." It seems probable, however, that Plato's idea of man may have arisen from a mal-observation of psychic phenomena. Thus, his idea of the spirit may have been derived from his observation of the operation of the subjective faculties; and his idea of the soul, from his observation of the objective mental activity, or *vice versa*. In other words, he failed to observe that the objective mind, instead of being an entity, is merely the function of the human brain, and necessarily ceases with the death of the body; whereas the subjective mind belongs to a distinct entity, which is apparently capable of sustaining an existence independently of the body. It is this entity which modern philosophers denominate the "soul" or the "spirit," — the two terms, in the vocabulary of modern spiritual philosophy, being generally synonymous. It is true that there are still to be found occasional representatives of the "telescopic" school of spiritual philosophy, who hold to the old doctrine of three entities, — body, soul, and spirit; but no one of them has ever been able to point to a single facts which discloses the existence of more than two. There are others who cling to the old vocabulary, but admit that there are but two entities, namely, the body and the soul; the spirit, in their philosophy, being the life principle which animates both body and soul, as well as all organic Nature. In this sense there is no objection to the phrase, save that it has a tendency to confuse the unscientific mind and to lead it into a maze of speculation which is as unprofitable as it is unscientific.

This apparent digression is made for the purpose of placing the conclusions derived from the speculative philosophy of the ancients in sharp contrast with the intuitional perceptions of the unperverted human soul, especially that of Jesus of Nazareth. I shall now follow it up by showing that the facts developed by modern scientific investigation confirm and emphasize the philosophy of Jesus in all its purity and simplicity. For it is a fact of most profound significance that Jesus, although he must be presumed to have been acquainted with all the philosophy of his day, never gave utterance to a word which indicated his belief in that complicated spiritual structure of man which has been so industriously proclaimed by both ancient and modern philosophers. He taught the simple truth that man has a soul; he demonstrated that truth by the production of phenomena of the soul; he promulgated the doctrine that the soul is capable of sustaining an existence independently of the body, and he taught mankind how to deserve and how to attain immortal life. This comprises all that man needs to know of spiritual philosophy, and it will not be denied that in its simplicity of statement it bears the impress of scientific truth. It is the province of science to ascertain whether observable facts confirm his doctrines.

Should a scientist who is unfamiliar with the recent developments of psychic science, be asked what would constitute conclusive evidence to his mind of the existence of a soul in man, he would doubtless reply that he must first be convinced that the mind does not follow the conditions of the body and brain. Then he would launch out into a more or less learned dissertation, reminding us that all human experience goes to show that as the body grows weaker the mind grows weaker; that a disease of the brain produces insanity or imbecility; that certain organs of the mind can be modified, inhibited, or totally destroyed by a surgical operation; that mechanical pressure upon the brain produces total unconsciousness and insensibility; that when the body dies, all manifestations of mind cease at once and forever, etc. In short, he would lead us through all the stock arguments of materialistic science which go to prove that the mind is not an entity, but a function of the physical brain, and that it necessarily ceases to manifest itself when the brain loses its vitality. All this we have heard before, *ad nauseam*, and all this we admit to be true, so far as the objective mind is concerned; but we hasten to remind him that the researches of modern science have developed the fact that man has a dual mind, — two planes of consciousness, — a normal and a super-normal plane, and that the latter manifests itself when the former is inhibited. In other words, when the brain is asleep and all the objective senses or faculties are in complete abeyance, the super-normal or subjective faculties are capable of intense activity. If he has heard of hypnotism as developed by orthodox scientists, he is ready to admit that man has a dual mind in the sense that it acts in one way under certain conditions of the body, and in another way under certain other conditions of the body. "It is the same mind," he adds, "its powers and functions being susceptible to modification either by peripheral stimuli or by the inhibition of activity in certain nerve centres and the stimulation of others to abnormal activity." All this has a very learned sound, in that it is characterized by the oracular indefiniteness of true materialistic science when dealing with problems beyond its legitimate domain. Should we then remind him that the facts of experimental psychology show that duality of mind means vastly more than is implied in his definition, — that, in fact, it is demonstrable that man possesses a dual mind, — he would doubtless inform us at once that such a thing is "impossible," that it is "contrary to the nature of things," that it is "subversive of all psychological science" with which he is acquainted, and, worst of all, "it would destroy the value and significance of all the learning that has been bestowed in times past upon the science of psychology;" in short, "that all the vast bibliography of the old psychology would have to be revised and rewritten"[1] if the dual-mind theory is demonstrable. Of course, all this is very shocking and subversive and revolutionary, and all that; but we will suppose our scientist to be a fair-minded man, and we will venture one more question, namely, what demonstration, short of pulling the two minds out of a man with a pair of forceps, weighing them in a balance, and carving them with a scalpel, would be considered adequate proof of the actual existence of two minds in man? I submit that the conditions to be prescribed by the most exacting scientist could be no more severe than the following: —

1. It must be shown that man possesses attributes and powers independent of each other and irreconcilable with each other except by the hypothesis that he is endowed with two minds.

2. That each is capable of independent action while the other is in complete abeyance.

3. That each must possess powers and limitations not possessed by the other.

4. That each must, in the normal man, perform functions which the other is incapable of exercising.

5. That one mind must normally be subordinate to the other.

6. That there must be some evidence of the survival of one after the extinction of the other.

7. That each of the foregoing propositions must be

[1] I am quoting from memory the words actually employed by an eminent scientist when confronted with the theory of duality of mind.

demonstrated by an appeal to observable facts that are susceptible of no other rational interpretation.

I think it will be conceded by the most sceptical that if the foregoing propositions can be fairly established, it will constitute at least *prima facie* evidence of the existence of a soul in mankind. When this is done, it will be followed by other considerations which will be demonstrative of that proposition. In the mean time, as the above propositions are nearly related, they will be considered together.

The broad line of distinction and demarcation between the two classes of attributes consists in the fact, which is of every-day observation and universal experience, that normally each class of attributes manifests itself while the other is quiescent. This fact is brought to universal consciousness in the phenomena of dreams. If dreams had but recently been brought to the attention of the civilized world, they would now be considered the most wonderful phenomena of the human mind. Being the common experience of all mankind, their real significance has been to a great extent overlooked. The reason for this is found in the fact that their most common form of manifestation can be traced either to peripheral stimuli, or to the uppermost waking thoughts of the sleeper. But it occasionally happens, and has happened throughout all the ages of which history or tradition gives us any account, that men have dreamed dreams which cannot be traced to anything within the known physical or mental environment of the dreamer. Dreams which give warning of impending danger; dreams which demonstrate the fact of communion with friends at a distance; dreams which solve problems far beyond the objective or normal capacity of the dreamer, — are among the phenomena which point clearly to a consciousness distinct from and independent of his normal consciousness, and possessing a power of perception of truth which reaches out far beyond the range of the objective senses. These are phenomena which have been observed in a haphazard, unintelligent way throughout all the ages, and would doubtless have continued to be so observed had not the phenomena of mesmerism or hypnotism been brought to the attention of science. Hypnotism enables us to study the phenomena of dreams by experimental reproduction; and in this sense it may be defined to be *the power to reproduce the phenomena of dreams.* It is that, but it is more. It is the power, not only to reproduce, but to control the phenomena, and carry them to an intelligible conclusion. Hypnotic phenomena possess all the salient characteristics of dream phenomena, and are governed by the same laws, modified only by methods of induction. The total or partial abeyance of the objective senses (sleep) is the first requisite in each ease. In hypnotism the subject is *en rapport* with the hypnotist, and his dreams are controlled by the suggestions of the latter. In natural sleep the subject is *en rapport* with himself, and his dreams are controlled sometimes by the suggestions conveyed in the current of his waking thoughts, and sometimes by those of peripheral stimuli. This is practically all that differentiates hypnotic sleep from natural sleep (Bernheim). When the sleep is induced by hypnotic processes, the subject may always be made to dream by making oral suggestions, and frequently by mere mental suggestion. He can also be made to dream by peripheral stimuli, such as applying heat or cold to his body, or by placing him in attitudes suggestive of certain mental emotions, or by causing music to be played in his presence.[1] There is another point where the phenomena of hypnotism and dreams exactly coincide which deserves particular attention. It is well known that, when

[1] See "Some Physiologic Effects of Music in Hypnotized Subjects," by Aldred S. Warthin, Ph.D., M.D., Demonstrator of Clinical Medicine in the Michigan University, Medical News, July 28, 1894 (Philadelphia).

sleep is profound, dreams are not remembered. Nevertheless, it was demonstrated many years ago, by the earlier psychologists, that, no matter how profound the sleep, dreams do not cease. It is precisely so in hypnotism. If the hypnosis is profound, the subject does not remember the experiences through which he has passed. It is, however, in that state of profound hypnosis that the most wonderful phenomena are produced. It is then that the evidences are most conclusive of a second intelligence existent in man; for it is then that the distinctive attributes and powers of the subjective entity come to the surface.

This, then, is the broad dividing line which separates the two hypothetical entities. It is so plain and palpable that it cannot fail to be observed by every one who has witnessed the phenomena of hypnotism; and it consists, as before remarked, in the fact that the essential condition precedent to the successful exercise of the distinctive attributes and powers of each of the two minds is that the other must be in a state of quiescence. This fact alone constitutes *prima facie* evidence of duality, at least in the limited sense of the term.

It will now be in order to ascertain the distinctive characteristics which differentiate the objective and subjective mental states, with the view of determining whether they point logically to the conclusion that the phenomena are the product of two distinct mental organizations.

The first distinctive characteristic of the subjective mind, which differentiates it from the objective in a most marked and unmistakable manner, consists in the fact that it is constantly amenable to control by the power of suggestion. This fact has been noted in a previous chapter, and is restated here for the sake of a symmetrical grouping of the whole of the leading characteristics of the subjective entity. It need not be dwelt upon at this time further than to remark upon its universality. The scientists to whom the world is forever indebted for the discovery of the law of suggestion, did not realize that it applied to other than the phenomena evoked by experimental hypnotism. Indeed, Bernheim himself expressly founds his definition of hypnotism upon the assumed fact that it only "increases the susceptibility to suggestion." In other words, he assumes that all men are more or less susceptible to suggestion "in their waking state." That this is true only in the limited sense that all men in their normal condition are more or less susceptible to persuasion or argument, is demonstrated by every experiment legitimately conducted. It is true that a very slight degree of hypnosis is required in very sensitive subjects, or in those who have often been hypnotized by the operator, as was the case in Bernheim's experiments; but that some degree of subjectivity or abeyance of the objective faculties is required to render a person susceptible to the power of suggestion proper, is all but self-evident. Thus it is a fact of common experience that when one goes to a dentist's office to have an aching tooth extracted, he invariably finds upon his arrival that the tooth has ceased to ache. In other words, Nature has hypnotized him so far as to produce anæsthesia in the refractory nerve, although he appears to be in a perfectly normal condition.[1] If, now, some one who understands the power of suggestion, and knows how to apply it, should make the suggestion to the patient that the anæsthesia would be continued until the tooth is extracted, no pain would be experienced from the operation.

The points to be here observed are, first, that it requires but a slight degree of hypnosis to render one susceptible to the power of suggestion; secondly, that some degree of abeyance of the objective faculties is necessary to render suggestion effective; thirdly, that the law of suggestion is

[1] See "Hypnotism a Universal Anæsthetic in Surgery," N. Y. Med. Jour., Dec. 22, 1894.

universal, and applies to all degrees of subjectivity and to all psychic phenomena of whatever name or nature; and fourthly, that the subjective mind is controllable by suggestion against reason, experience, and the evidences of the senses. It is unnecessary to remark that the objective mind, in its normal state, is not susceptible to such control.

It will not be denied that this difference between the two mental states of man possesses a veritable significance of some kind, nor can it be doubted that the solution of the problem will prove of the utmost importance to the human family.

The second proposition bearing upon the subject is that the subjective mind is incapable of reasoning inductively. This will readily be seen to be a corollary of the law of suggestion; that is to say, under the law of suggestion the subjective mind, when it reasons at all, necessarily takes its premises from the objective mind. In experimental hypnotism its premises are the suggestions of the hypnotist. From those suggestions it will reason deductively with most marvellous acumen. Indeed, its power of correct deduction from any premises suggested seems to be practically perfect. And this is equally true whether the premises are true or false. Like the ancient Greek philosophers, it does not disdain facts when they will serve its purpose; that is, when they serve to sustain the proposition embraced in the suggestion. Indeed, it will marshal all such facts that are within the storehouse of its perfect memory; but it will persistently ignore all facts which militate against the suggestion. It is impossible that it should do otherwise. The inexorable law of suggestion interposes an insuperable barrier against independent thought, which is the very essence of induction. The subjective mind cannot marshal facts except on one side of a question, and that side is in favor of the suggestion which happens for the time being to be uppermost. That is as true of the subjective mental operations of every-day life as it is of experimental hypnotism.

The third proposition is that the power of the subjective mind to reason deductively from given premises is practically perfect. This has been mentioned in the foregoing paragraphs, but it requires further elucidation. I do not set this down as a distinctive characteristic of the subjective mind which is antithetical to the objective; for it is obvious that the latter possesses that power. But it is worthy of note, for the reason that the difference in degree is so marked that it practically amounts to a distinctive attribute of the subjective mind. Nor do I assert that its power of correct deduction is perfect. It is manifestly impossible to know just when the conditions are perfect for the manifestation of the highest powers. But that under favorable conditions it is practically perfect, does not admit of doubt in the minds of those who have intelligently observed the phenomena of experimental hypnotism, or, indeed, any of the higher phases of psychic phenomena. When this prodigious power of correct deduction is properly understood and appreciated, it will be found to furnish the key to many obscure problems in psychic science. As a single instance I will mention the fact that many of the phenomena of so-called prevision, or prophecy, may be traced directly to the power of correct deduction from premises derived from any of the myriad sources from which a suggestion may be imparted or knowledge obtained. Telepathy, for instance, is one of the sources of information through which the subjective mind obtains knowledge of facts not consciously possessed by the objective intelligence of either of the parties concerned. It is safe to say that most if not all of the mysterious cases of prevision may be traced to this cause. I do not say that the subjective mind of man may not possess the inherent power of correct prevision independently of knowledge of the subject-matter or the processes of reasoning. I do not know. Many stories are current which would seem to indicate the existence of that power. But such stories are usually very far from being sufficiently well authenticated to warrant any scientist in giving them anything more than a provisional place in psychic science. Certain it is that all the scientifically verified accounts of correct prevision may be traced directly to the soul's wonderful power of correct deduction; and all that is mysterious regarding its sources of information may be traced to telepathy and perfect memory. When these two powers are taken into consideration, it will readily be seen that the subjective mind is in possession of sources of data of which the objective intelligence of mankind has as yet but a faint conception; and when to these sources of information is added the power of perfect deduction, it will be seen that much, if not all, of that which has seemed mysterious and inexplicable except by reference to supermundane sources of information, is easily explained by reference to natural laws with which the world is fast becoming acquainted.

May not this be the fountain of "this pleasing hope, this fond desire, this longing after immortality"? May it not be the origin of that emotion of the human mind which has been designated as an "intuitive knowledge of a life to come"? May not the soul have experiences of spiritual life so far removed from the physical realm that no cognizance of them can be taken by the objective consciousness? May not the soul be able to reach a state of consciousness so exalted as to enable it to come into contact and conscious communion with a deific intelligence, and to imbibe truth from its Eternal Source? May not the soul reach such a state of illumination during the unconscious moments of the physical senses that all spiritual truth will be open to its intuitive perceptions? May not the soul, either in its state of induced exaltation or by the exercise of its normal powers, be able to take cognizance of spiritual facts from which it may, by the exercise of its marvellous powers of correct deduction, be able to demonstrate to its own consciousness the fact of immortality? All these questions, and more, will be asked by the earnest seeker after tangible evidences of a future life. They are interesting, if not pertinent, questions; and were we indulging in the pleasing phantasies of speculative philosophy, and could thus afford to dispense with facts, we might construct an argument for immortality that to many would seem impregnable. But there is one insuperable obstacle in the way which must forever prevent the construction of a conclusive argument based upon these hypothetical powers. The inexorable law of suggestion interposes itself at the very threshold of the argument, and casts a doubt upon the verity of the premises. It might even be demonstrated that the soul's power of correct deduction from given premises was perfect and infallible; yet, when the correctness of the premises is in doubt, the argument based upon them is necessarily invalid. In other words, the soul, so long as it inhabits the body, is never exempt from the operation of the law of suggestion. Hence it is often impossible to know whether its supposed perceptions are veridical or are merely subjective hallucinations resulting from auto-suggestion or from a suggestion imparted to it from some extraneous source. It is evident, therefore, that we must look elsewhere than in hypothetical perceptions or intuitions, unsupported by demonstrative evidence of their verity, for logical proof of a future life. I have dwelt thus far upon the subject of the deductive power of the subjective mind, not because it differentiates the one mind from the other, except in degree, but because of its general interest in that it furnishes an explanation of much of the phenomena of so-called prevision. In the ensuing chapter I will return to the consideration of those faculties of the soul which present distinctive points of difference from the faculties and functions of the objective mind.

CHAPTER XIII.

HAS MAN A SOUL? (*continued*).

The Perfect Memory of the Subjective Mind. — Memory and Recollection Differentiated. — Sir William Hamilton's Views. — Intuitional Powers of Perception of Nature's Laws. — The Seat of the Emotions. — The Three Normal Functions of the Subjective Mind. — The Infant's Development from Savagery to Civilization. — Total Depravity. — Dangers of Subjective Control. — Telepathy a purely Subjective Faculty. — Abnormality of Psychic Manifestations. — Ill Health a Condition precedent to their Production. — They grow Stronger as the Body grows Weaker. — Strongest in the Hour of Death. — The Objective Mind perishes with the Brain.

THE fourth characteristic of the subjective mind, which distinguishes it from the objective, consists in the fact that the former is endowed with a perfect memory. In saying this, I am not unmindful of the fact that the objective mind is also endowed with a memory; but its manifestations are feeble in comparison with the prodigious power of the subjective mind. Properly speaking, the difference between the two would be defined by the employment of the word "memory" to designate the faculty in the subjective intelligence, and the word "recollection" to designate the corresponding faculty in the objective mind. Memory, in this sense, is the actual and distinct retention of recognition of past ideas in the mind (Webster). Recollection is the power of recalling ideas to the mind; in other words, it is the power of re-collecting the ideas which have once been in the mind, but are, for the time being, forgotten. The latter faculty varies in strength in different individuals. Subjective memory is the absolute retention of all ideas, however superficially they may have been impressed upon the objective mind; and it admits of no variation in power in different individuals. It must not be understood that all manifestations of subjective memory are equally perfect. That is obviously impossible, for the reason that subjective conditions are not always perfect; but experimental hypnotism develops the fact that subjective memory is exalted, other things being equal, just in proportion to the depth of the hypnosis.

The German psychologists noted this phenomenon many years before the English philosophers took it into account; and it was not until Sir William Hamilton brought it to the attention of the English-speaking public that it was seriously considered as a factor in psychological science. Sir William designated it as "mental latency;" and he went so far as to hold that all recollection consisted in rescuing from the storehouse of latent memory some part of its treasure. His hypothesis necessarily presupposed latent memory to be perfect, and he cites many cases in support of that supposition. The curious part of his hypothesis, however, consists in the fact that whilst he considers it a normal mental process to elevate a part of the latent treasures of the mind above the threshold of consciousness, he recognizes the fact that it is only under the most intensely abnormal conditions that the whole content of the magazine of latent intelligence can be brought to light. He says: —

"The second degree of latency exists when the mind contains certain systems of knowledge or certain habits of action which it is wholly unconscious of possessing in its ordinary state, but which are revealed to consciousness in certain extraordinary exaltations of its powers. The evidence on this point shows that the mind frequently contains whole systems of knowledge which, though in our normal state they may have faded into absolute oblivion, may in certain abnormal states, as madness, febrile delirium, somnambulism, catalepsy, etc., flash into luminous consciousness, and even throw into the shade of unconsciousness those other systems by which they had for a long period been eclipsed and even extinguished. For example, there are cases in which the extinct memory of whole languages was suddenly restored, and, what is even still more remarkable, in which the faculty was exhibited of accurately repeating, in known or unknown tongues, passages which were never within the grasp of conscious memory in the normal state. This degree, this phenomenon of latency, is one of the most marvellous in the whole compass of philosophy." [1]

He then cites some most remarkable instances demonstrative of the perfection of subjective memory.[2]

It is obvious that Sir William had not studied the phenomena of experimental hypnotism, or he would have discovered many facts which his hypothesis of mental latency could not account for. Amongst others he would have discovered that physical disease of a very pronounced character is not essential to the production of phenomena exhibiting the marvellous perfection of subjective memory; and that an hypnotic subject can be so trained that, even in an apparently normal condition, he can be caused to memorize a whole page of printed matter by gazing upon it but two seconds of time.[3] He would have found in the dual hypothesis a complete explanation of the facts which he labored in vain to explain, and a more direct road to a demonstration of what he labored so assiduously to prove.

The fifth faculty of the subjective mind, which distinguishes it from the objective intelligence, consists in its power under certain conditions, not yet clearly defined, of apprehending by perception or intuition, and without the

[1] Lectures on Metaphysics, p. 236.

[2] For a fuller discussion of this subject, see "The Law of Psychic Phenomena," ch. iv., v. See also Beasley on the Mind; Abercrombie on the Intellectual Powers; and Coleridge's Biographia Literaria.

[3] See Bertolacci's Christian Spiritualism, p. 30.

aid of the process of induction, the laws of Nature. As this branch of the subject has been treated in a former chapter, it is mentioned here merely for the sake of symmetrical grouping.

The sixth distinctive characteristic of the subjective mind consists in the fact that it is the seat of the emotions. In that alone exists the emotional element in man. The objective mind is pure intellect, — cold, deliberate, reasoning. During the normal physical life of man it is the dominating power in the dual mental organization. This is necessarily true for the reasons, first, that it would be impossible for the two entities to maintain harmonious relations if one were not normally subordinated to the other; and second, that the dominant power must be that which is endowed with the faculty of reasoning by all processes and in all directions. That power is the objective mind, and it is enabled to maintain its ascendency solely by virtue of the fact that the subjective mind is normally amenable to control by the power of suggestion. A complete reversal of the order — that is, the subjective mind in control — is what constitutes insanity. A partial reversal constitutes partial insanity, and is also the source of all vice and immorality. Indeed, vice, in this sense, is a form of insanity; that is to say, the same cause operates to produce both, the difference being in degree only.

Before proceeding, however, to discuss this branch of the subject, it is logically necessary to verify the fundamental proposition; namely, that the subjective mind or soul is the seat of the emotions. A few words will be sufficient for this purpose.

It will not be denied that what we call "instinct" in animals is a purely subjective endowment. Its acts are performed independently of objective reason or intelligence.

"Instinctive acts, so far as the individual exhibiting them is concerned, are not the result of instruction or experience. This is one of the most prominent points wherein the actions in question differ from those which proceed from intelligence and reason, performed for a definite purpose. These latter are necessarily due to impressions conveyed to the mind through the senses and nerves, and are, therefore, of eccentric origin. The former are prompted by a force acting altogether without the agency of intelligential external sensations of any kind, and are of internal origin."[1]

It may be defined as follows: —

Instinct is that innate faculty of the subjective mind, which all organic beings possess, by which they are impelled to perform certain volitional acts without being prompted thereto by the objective intellect, which acts are preservative of the well being or life of the individual, or of the species to which it belongs.[2]

There are three clearly defined instinctive emotions, two of which are common to all organic beings. They are: 1. The instinct of self-preservation; 2. The instinct of reproduction; 3. The instinct which impels the parent to acts preservative of the well being or life of the offspring. To the last must be added that which is common to man and a few of the animal creation; namely, that instinct which impels the individual to acts preservative of the lives of the members of the species to which it belongs, without regard to age or consanguinity. This may be regarded as a higher instinct; but it is obviously the result of social or political organization, and is merely a modification of the instinct of self-preservation. To these three instinctive emotions of the human mind should be added a fourth, which is no less clearly defined, and which is well-nigh universal; namely, the instinct of religious worship. This, however, does not pertain directly to the subject under consideration. It is, however, necessarily a subjective emo-

[1] Dr. William A. Hammond's Treatise on Insanity, p. 137.

[2] This is a slight modification of Dr. Hammond's clear and comprehensive definition.

tion, as it pertains exclusively to the human soul. We have, then, three distinct primary emotions of the mind which are obviously and necessarily attributes of the subjective mind, as distinguished from the objective intelligence. In man they all pertain to the perpetuation of the human race, and their resultant acts constitute the normal functions of the human soul. Moreover, they are the only normal functions of the soul in its relations to the physical organization.

Leaving out of present consideration the emotion of religious worship, all human emotions are traceable to one or the other of the primary instinctive emotions just mentioned; that is to say, all emotions are modifications or combinations of the primary emotions. This being true, it follows that my original proposition is true; namely, that the subjective mind or soul is the seat of the emotions.

Now, inasmuch as all immorality, vice, and crime are perversions of the human emotions or passions, it follows that immorality, vice, and crime are the results of giving to the subjective mind undue control of the dual mental organization. In other words, Reason abdicates her rightful authority and power, and the subjective mind usurps its place, but without having the ability to perform its functions.

That this is true, is evidenced by every phenomenon of physical and mental life and growth. Thus, the history of the life of every human being from infancy to adult age, is an epitome of the history of the development of the race from savagery to civilization. The newly born infant is purely subjective; its objective mind is a blank. It is then governed alone by instinct. It is an animal, with nothing in its mental development to distinguish it from the cub of the bear or the lion's whelp, except its physical conformation and its more absolute helplessness. It quickly develops from the purely animal existence, but the first step in its evolution is to a state of savagery. Its instincts are wholly selfish, it is destitute of conscience, and totally oblivious of the rights of others. Its next step is to a state of barbarism. That "the small boy is a barbarian" is a vulgar statement of a scientific truth; and the saying belongs to the vocabulary of Evolution. From barbarism he soon emerges into semi-barbarism, then to civilization, and finally to enlightenment. The point to be noted is that the various steps in advance are due wholly to objective education. Modified to a certain extent by heredity, the degree of a child's progress from infancy to manhood, or rather from savagery toward enlightenment, is determined by its environment; that is, by the character of its objective education. In other words, just in proportion to the excellence of the mental and moral training of the objective mind, and just in proportion to its dominating power over the subjective faculties, will be the excellence of the moral character of the individual. If this training has been good, and if objective reason has exercised its legitimate control over the subjective passions or propensities, — in short, if normal conditions are allowed to prevail, a perfect mentality and a lofty moral altitude will be reached. But if, for want of proper training, the subjective mind is allowed to maintain in adult life that ascendency which is its normal condition in infancy, moral degradation, if not crime or insanity, is sure to result. The doctrine which affirmed the "total depravity" of the "natural man" derived its origin from a constant observation of these phenomena of the human mind; and, in the absence of all knowledge of the true relationship existing between the two minds, of their respective and relative powers, functions, and limitations, who shall say that the conclusions of the ancient philosophers were not justified? The control of the subjective mind by the objective, through the power of suggestion, is normally absolute and perfect when that power is asserted and maintained. Nevertheless, there appears to come a time in every man's life when there is a conflict of jurisdiction, — a struggle for ascendency between the natural passions and instinctive propensities on the one hand, and the restraints of reason, morality, and religion on the other. Saint Paul, in graphic phrases, gives expression to that inward conflict when he says: —

> "I find then a law, that, when I would do good, evil is present with me. For I delight in the law of God after the inward man: but I see another law in my members, warring against the law of my mind, and bringing me into captivity to the law of sin which is in my members. O wretched man that I am! who shall deliver me from the body of this death?"

Saint Paul's experience is by no means unique; but the remedy is in the hands of every normal man; and that is to assert and maintain the ascendency of objective reason over the instinctive emotions. Not that there is anything inherently sinful or abnormal in their legitimate indulgence, for there is not. Like every other faculty of the human mind and soul, they have their legitimate sphere of operation; and their normal exercise is beneficent. Without it the world would be depopulated in a generation. With it was organic evolution made possible. The danger lies, not so much in sporadic cases of predominant passion in the otherwise normal man; but there is an appalling and constantly growing danger in the modern tendency towards the cultivation of subjective powers in utter ignorance of the fundamental laws which pertain to their operation. It is a danger which in its tendencies threatens the foundations of civil society; for it presents itself in such insidious guise that the innocent alike with the guilty are in danger of being drawn within its vortex. This, however, is a branch of the subject which must be treated in a future chapter, and I will therefore no longer digress.

The seventh faculty possessed exclusively by the subjective entity consists of the power to move ponderable objects without physical contact. In this it is meant to include all the physical phenomena of so-called spirit manifestations. I need not dwell upon this branch of the subject, for the phenomena are so thoroughly attested that it would be a waste of time to attempt to convince those who mistake ignorance for scepticism. Besides, to those who agree with me in ascribing these manifestations to the agency of the psychic who produces the phenomena, it is self-evident that it is by the subjective entity that they are produced. All others will agree that they are not due to any known physical power. Moreover, the subject has been so fully treated in "The Law of Psychic Phenomena" that it could not be enlarged upon here without unseemly repetition.

The eighth faculty or power of the subjective mind which clearly and sharply differentiates it from the objective mind, is that of telepathy, or the power of one mind to communicate intelligence to another, otherwise than through the recognized channels of the senses. That this power exists, is no longer a matter of doubt in well-informed scientific circles. It was demonstrated many years ago by the old mesmerists who succeeded Mesmer; and if any reasonable doubt existed after their demonstrations, it has within the last decade been dispelled by the carefully conducted scientific experiments of the Society for Psychical Research. That Society, in addition to its regular "Proceedings," a large volume of which appears every year, has caused to be published two large volumes, aggregating over thirteen hundred pages, entitled "Phantasms of the Living," which are filled with demonstrative evidence of the existence of the power of telepathy. This work, however, is now out of print; but it has recently been ably supplemented by a small volume by Frank Podmore, one of the secretaries of the Society for Psychical Research, which gives, in concise form, a *résumé* of the evidence which has been collected by the Society since its organization in 1882. In fact, the evidence is accumulating in every intelligent household in the civilized world. Circumstances innumerable, which were in former times passed by as curious coincidences, or were ascribed to supermundane agency, are now intelligently observed and referred to their proper source, since science has rescued the phenomena from the domain of superstition.

At first it was supposed that the phenomenon pertained solely to the objective mind, and that what the agent was consciously thinking of was necessarily that which was conveyed to the mind of the percipient. But that theory was soon abandoned in view of constantly occurring phenomena which could not be thus explained. Thus it was found that thoughts were transmitted which were not consciously in the agent's mind; and that, as Podmore observes, —

> "The idea can be transferred from the sub-conscious to the sub-conscious; and indeed there is some ground for thinking that, outside of direct experiment, the intervention of the conscious [objective] mind in the telepathic transmission of thought is exceptional. Even in some of the most striking experimental cases it has been shown that either agent or percipient, or both, were asleep or entranced at the time."[1]

These conclusions, although expressed with the caution of the true scientist, are obviously correct. In telepathic experiments by means of hypnotism, the subjective mind of the percipient is alone concerned. Indeed, all the evidence on the subject goes to prove that telepathy is a power belonging exclusively to the subjective mind; and that in the spontaneous exercise of that power it is by mere accident that the objective mind participates in, or is cognizant of, either the transmission or the reception of the communication. That is to say, it is quite evident that telepathic communion is very common, if not constant, between mem-

[1] Apparitions and Thought-Transference, p. 391.

bers of the same family, or those who have a vital interest in each other's welfare; although it is comparatively rare that the content of the communication is elevated above the threshold of normal or objective consciousness of either the agent or the percipient. As I have pointed out in the chapters on spiritism, it is to the fact that telepathy is purely a subjective power that all the seeming mystery attached to so-called spirit communications is to be attributed.

It is clear, therefore, that the power of telepathy has nothing in common with objective methods of communications between mind and mind; and that it is not the product of muscle or nerve or any physiological combination whatever, but rather sets these at naught, with their implications of space and time.

"It is a quality that defies distance, is instantaneous, is not dependent on terrestrial states, is most apparent in our least conscious moods and in our least wakeful hours, is strongest in the undeveloped intellectually, is conspicuous in the moments when organization is dissolving, in the hour of death, — is certainly as near to our conception of soul as a thing can be."[1]

The ninth characteristic of the subjective entity which clearly differentiates it from the objective mind, in power, function, and attribute, consists of the fact that its activity and power are in inverse proportion to the vigor of the body. This is the most important of all of the distinctive differences between the two minds or intelligences, for it is not only a strong argument for the existence in man of a distinct entity, but it goes far towards proving that this entity is capable of sustaining an existence independently of the body. *If a man has a power that transcends the senses, it is at least presumptive evidence that it does not perish when the senses are extinguished.*

That the activity and power of the subjective mind is in inverse proportion to that of the body, is evidenced by

[1] O. B Frothingham, in Harper's Magazine, August, 1860, p. 205.

every phenomenon of subjective mental action. Beginning with the simplest hypnotic experiment upon a healthy subject, in the first stage of subjective activity the physical condition of the patient can hardly be distinguished from the normal. Deepen the hypnosis, and the subjective manifestations will increase in power and intensity. Continue the process until a hypnotic lethargy is induced, and the manifestations will continue to grow stronger in proportion. This must be understood as a general statement of a condition which, within certain limits, varies with each individual. There are, however, many psychics whose strongest manifestations are produced while the body is in an apparently normal condition. In fact, no general rule can be laid down which will apply to all cases, except this, that, the longer and more persistently the production of psychic phenemena is followed up, the weaker will become the physical organism of the psychic. But as this branch of the subject will be treated in a future chapter, I will come direct to the salient point to which I wish to invite attention. It is this: When disease seizes the physical frame and the body grows feeble, the objective mind invariably grows correspondingly weak. Not so the subjective mind; for, as the body grows weak, the subjective mind grows strong, and it is strongest in the hour of death. Indeed, when death approaches, no matter what form it assumes, the moment its inevitability is realized, it is no longer feared, and pain ceases. At that supreme moment the subjective mind takes complete possession, the objective senses are benumbed, the body is anæsthetized, and the patient dies, "without pain and without regret" (Hammond). In the mean time, as the objective mind ceases to perform its functions, the subjective mind is most active and powerful. The individual may never before have exhibited any psychic power, and may never have consciously produced any psychic phenomena; yet at the supreme moment his soul is in active communion with loved ones at a distance, and the death message is often, when psychic conditions are favorable, consciously received. The records of telepathy demonstrate this proposition. Nay, more; they may be cited to show that in the hour of death the soul is capable of projecting a phantasm of such strength and objectivity that it may be an object of sensorial experience to those for whom it is intended. Moreover, it has happened that telepathic messages have been sent by the dying, at the moment of dissolution, giving all the particulars of the tragedy, when the death was caused by an unexpected blow which crushed the skull of the victim. It is obvious that in such a case it is impossible that the objective mind could have participated in the transaction. The evidence is, indeed, overwhelming, that, no matter what form death may assume, whether caused by lingering disease, old age, or violence, the subjective mind is never weakened by its approach or its presence. On the other hand, that the objective mind weakens with the body and perishes with the brain, is a fact confirmed by every-day observation and universal experience.

CHAPTER XIV.

HAS MAN A SOUL? (*continued*).

Recapitulation. — A *Prima Facie* Case. — Concurrent and Antagonistic Hypotheses. — The Law of Suggestion. — A Case of "Mediumistic" Development. — The Alleged Spirit Control assumes a Dictatorship. — It develops a Passion for Music. — Music the Language of the Emotions. — A purely Subjective Faculty. — Subjective Music and Objective Music Differentiated. — The Dual-Mind Theory. — Absurdities Involved in the Single-Mind Theory.

IT must now be provisionally assumed that it has been proven that the subjective mind is endowed with powers, and circumscribed by limitations, which clearly differentiate it from the objective mind. For convenience of reference and facility of recollection, the following recapitulation is presented: —

1. The subjective mind is constantly amenable to control by the power of suggestion.

2. It is incapable of independent reasoning by the processes of induction.

3. Its power to reason deductively from given premises to correct conclusions is practically perfect.

4. It is endowed with a perfect memory.

5. It is the seat of the emotions.

6. It possesses the power to move ponderable objects without physical contact.

7. It has the power to communicate and receive intelligence otherwise than through the recognized channels of the senses.

8. Its activity and power are inversely proportionate to the vigor and healthfulness of the physical organism.

9. It is endowed with the faculties of instinct and intuition, and, under certain conditions, with the power of intuitive cognition or perception of the laws of Nature.

It needs no argument or illustration to show that the objective mind has little in common with the subjective in any of the foregoing attributes, powers, and limitations. The objective mind (1) is manifestly not controllable by the power of suggestion in the sense in which the subjective mind is so controlled, — that is, against reason, experience, and the evidence of the senses; 2. It is capable of inductive reasoning; 3. Its power of deductive reasoning is by no means perfect, nor does it approach perfection; 4. Its memory, in its best estate, is very defective, and, comparatively speaking, amounts to nothing more than an uncertain, evanescent ability to recall a few of the more prominent ideas and impressions which it has once experienced; 5. It is absolutely destitute of emotion; 6. It cannot exercise the slightest kinetic force beyond the range of physical contact; 7. It is destitute of any power remotely akin to telepathy; 8. The essential prerequisite to the successful exercise of its highest powers and functions is a perfectly sound, healthy, normal physical organism; 9. It is endowed with no power which is remotely akin to instinct or intuition.

I submit that the mental characteristics of no two individuals ever presented a more violent contrast than exists between the objective and subjective minds of the human entity, in all their essential powers, functions, and limitations. I might claim the logical right to rest my case at this point; for it must be remembered that I have thus far sought not to prove the immortality of the soul, but to demonstrate the fact that man *has* a soul. In other words, I have merely sought to prove that which I have for convenience designated the "subjective mind" is in reality the mind of a distinct entity, and not merely one set or series of faculties which perform their functions under one condition of the body, whilst another set of faculties perform their functions under other bodily conditions. I say I might claim the logical right to rest my case here; for I submit that the bare statement of the facts which differentiate the two minds constitutes *prima facie* evidence that they belong to two distinct entities. The *onus probandi*, therefore, rests with those who hold the materialistic hypothesis, that man is a soulless being, possessing no attributes or powers that cannot be accounted for by reference to cerebral anatomy and physiology. I do not, however, intend to stop here, but will now proceed to show that there is no way of rationally accounting for the facts other than to predicate the actual existence in mankind of an entity which, in the vocabulary of spiritual philosophy, is denominated the soul.

The nature of the question, broadly speaking, admits of but two hypotheses. One is that the facts presuppose the existence of two separate and distinct minds, belonging to two separate, or separable, entities; and the other is that there is but one mind having two distinct planes of consciousness, or two sets of faculties. One or the other of these hypotheses is the true one. They cannot both be true, and yet, for the purpose of demonstrating the immortality of the soul, it is a matter of indifference which of the two is adopted: whether we consider man as having two distinct minds, or as having one mind which manifests certain attributes and powers under certain conditions, and other attributes and powers under certain other conditions, provided only that the crucial fact remains that certain of those powers and functions do not pertain to this life. In other words, it is a matter of indifference whether we employ the words "dual mind," or "two minds," or "two sets of faculties;" for the same logical result follows, whatever terminology may be employed in the discussion of the broad and pregnant fact that two sets of faculties exist in man, each possessing independent powers, functions, and limitations. Facts are independent of hypotheses. *Facts are primordial. Hypothesis is an instrument of logic for the scientific investigation of facts.* Hypotheses, considered in their relations to each other, are divisible into two classes; namely, *concurrent* and *antagonistic.* Concurrent hypotheses are those of which the ultimate conclusions coincide. Antagonistic hypotheses are those of which the ultimate conclusions are variant. It is often a matter of indifference which of two concurrent hypotheses is the correct one; and it is often impossible to ascertain with certainty which is scientifically correct. It is, however, generally ascertained, sooner or later, by the failure of one to explain facts collateral to the main question, but resident within its purview. When that occurs, the true scientist will immediately resort to the other, providing that one explains all the facts.

The hypotheses of duality of mind on the one hand, and of unitary mind with two sets of faculties on the other, are illustrations of concurrent hypotheses, inasmuch as their ultimate conclusions regarding a future life are identical; that is to say, they are each founded upon the one fact that man possesses subjective faculties that perform no normal function in physical life, and objective faculties which can perform no function in spiritual life. The conclusions are necessarily identical; namely, that faculties which perform no normal functions in this life must necessarily belong to a future life. Hence I have remarked, here and elsewhere,[1] that the dual theory is not a necessary premise to enable us to arrive at correct ultimate conclusions. That theory, however, will be constantly advanced, partly for the sake of clearness of statement, but principally because it is firmly believed to be scientifically correct. It must be borne in

[1] See "The Law of Psychic Phenomena," ch. i.

mind, however, that one mind with two sets of faculties is virtual duality; especially if it can be shown by reference to anatomy, as well as by experimental surgery, that there are actually existent two organs of mind. This I shall proceed to show in the succeeding chapter. In the mean time the dual hypothesis certainly explains all the facts, — that is to say, everything happens just as though man possessed a dual mind; and that is all that can be required of a working hypothesis.

The spiritistic hypothesis as opposed to the telepathic is a fair example of antagonistic hypotheses; for the conclusions are variant, and so are the hypotheses. I speak of spiritism as an "hypothesis," although it is hardly deserving of that designation; for it is, in fact, merely a short and easy method of avoiding, rather than promoting, a truly scientific investigation of the subject-matter. But, such as it is, it is an exceedingly convenient hypothesis; and, for the use of those who imagine that simplicity is a test of truth, it is admirably adapted. It is a summary way of disposing of an intricate problem, — of avoiding a difficulty instead of investigating its cause. It should be constantly borne in mind that an hypothesis is not a final dogma. It is, as before remarked, merely an instrument of investigation. To establish it as a scientific truth, there are two prerequisites: 1. It must explain all the facts; 2. There must be no other hypothesis capable of explaining all the facts. When these conditions are present, then, and not till then, is an hypothesis elevated to the dignity of a scientific truth, an established principle, a theorem.

I have made the foregoing remarks, at some risk of repetition, for the reason that I desire the reader to keep constantly in mind the fundamental principles of scientific investigation, and for the further reason that I desire him to measure the value of my arguments, past and future, by the highest known standards of scientific inquiry.

We will now briefly consider the nine propositions which set forth the salient characteristics of the subjective mind, and which seem to warrant us in postulating the dual-mind hypothesis.

The first in the order of statement as well as of importance relates to the law of suggestion. When that law was first discovered, its full import was not realized. As I have heretofore pointed out, it was discovered in the course of scientific experiments on hypnotic subjects, and it was, consequently, regarded as a law pertaining exclusively to experimental hypnotism. As long as the law was supposed to be confined to that narrow field of operation, the dual hypothesis was unnecessary; for most of the phenomena of experimental hypnotism could be accounted for on the theory that in the artificially induced hypnotic state the mind of the subject could be dominated by the suggestions of the hypnotist. But when it was discovered that the law of suggestion is the dominating principle which gives character and direction to all psychic phenomena, it was found that the single-mind theory was inadequate. And this inadequacy became more marked when it was discovered that a psychic can control his own subjective manifestations by exercising the power now known and recognized in the scientific world as "auto-suggestion." When this power is fully realized, it will be found to be demonstrative of the dual hypothesis, in that it fully and easily explains all the facts; whereas the single-mind theory is adequate to explain but very few, and the few that it can account for are of the least possible significance. As a single illustrative instance where the single-mind theory fails, it may be remarked that it has again and again been demonstrated that, by persistent auto-suggestion, the objective mind can control the subjective in direct contravention, not only to the beliefs, but to the positive knowledge of the former. Again, it is well known that, under the influence of suggestion, the subjective mind will not only assume control, but will act in direct opposition to the volition of the objective mind. An illustrative case has been brought to my attention as I write. A gentleman of this city had been induced to join a "developing circle" of spiritists who desired to establish a direct and independent line of communication with the other world. He very soon developed the power of automatic writing, and some very remarkable results ensued. As might have been expected, the power which moved his hand to write told him many things which he did not objectively know, and gave many remarkable exhibitions of power which constituted conclusive evidence to his mind of spirit identity and of the truth of the spiritistic hypothesis. But the remarkable feature of the case was that it soon began to assume a sort of dictatorship over his daily conduct, and on lines which he had least reason to anticipate. For instance, although he was not particularly devoted to music, yet his "control" insisted on being taken to concerts, oftentimes to his great inconvenience. When the "control" desired any indulgence, the gentleman was apprised of the fact by feeling a decided sensation in his right arm and hand, which sensation the gentleman soon learned to recognize as evincing a desire on the part of his "control" to communicate in writing. He often felt the impulse while walking the streets, and, upon being furnished a tablet and pencil, the "control" would direct him to go, perhaps to an adjoining street, to listen to a hand-organ. Sometimes it would ask for indulgences which were decidedly out of harmony with the settled moral principles of the gentleman's life; and so persistent were the requests that he was finally obliged to refuse to allow any communication in writing except on condition that those subjects should be tabooed. Fortunately the gentleman had not gone so far as to cease to desire to resist the subjective impulses; and he saw his danger in time to avoid its consequences. He had a superficial acquaintance with the law of suggestion and the theory of a subjective personality, but could not harmonize his experiences with that hypothesis. "The 'control' acted with perfect independence," he reasoned, "and frequently in decided opposition to my wishes, — that is, to the suggestions of my objective mind. How could that occur if the law of suggestion controls the subjective mind?" The answer is easy and perfect. The first and ever dominant suggestion in his mind was that the "control" was a spirit from the other world. His whole environment, during the time of the development of the power of automatic writing, was such as to force that suggestion upon his mind. He was in a "circle" organized for the purpose of obtaining communications from spirits of the dead; and all the communications were received under that suggestion. The "control," therefore, was assumed from the start to be an extraneous, independent, dominant, irresponsible power, which the medium could in no wise control or direct. That was the primary "suggestion" under which the "control" acted; and it necessarily carried out that suggestion by acting the part assigned to it. Of course, the idea never occurred to the gentleman that the very fact that the so-called spirit insisted on being taken to a concert was demonstrative that the intelligence was his own, and was inseparable from his physical organism for the time being. If it had been a foreign spirit possessing the powers ascribed to such spirits, it would seem that it could have heard the concert without the necessity of making a draft upon the gentleman's purse or his time. Be this as it may, the very vocabulary of spiritistic circles conveys the suggestion that the communicating intelligence is in "control" of the medium. If it did not act accordingly, and assume the right to control the medium, often in opposition to subsequent suggestions, the law of suggestion would have no place in the science of experimental psychology.

I have, of course, assumed that the "control" in this case was the gentleman's own subjective mind, and that the phenomena were produced in strict accordance with the laws which govern all subjective mental activity. Indeed, the phenomena are illustrative of those laws and principles in more than one sense of the word. As we have already seen, they illustrate one of the most subtle and intricate phases of the operation of the universal law of suggestion. They also incidentally illustrate a phase of subjective mental characteristics to which as yet little attention has been given by students of experimental psychology; namely, the power of music over the subjective mind. Some attention was given to it in my former work;[1] and Dr. Aldred S. Warthin, Ph. D., M. D., of the Michigan University, has made some very interesting experiments — as yet incomplete, however — illustrative of "Some Physiologic Effects of Music in Hypnotized Subjects," to which I have alluded in a former chapter. Chomet, a French author, has also written a work entitled "Effets et Influence de la Musique sur la Santé et sur la Maladie;" and Vigna, an Italian, has given us "Sull' importanza fisiologica e terapeutica della Musica." Little advance, however, has been made in the study of the physiological effects of music, although there is no doubt that the field offers rich results to the student who will give patient and intelligent attention to that line of experimental investigation. It will probably be found, however, that the physiologic effects are due to reflex action of the emotions upon the physical system, and not to any direct vibratory action upon the nerves. It should therefore first be studied as a psychological problem.

Music has been loosely described, by those who recognize its subjective origin, as "a passion of the human soul."

[1] See "The Law of Psychic Phenomena," ch. vi.

This is not scientifically correct. It may properly be described, however, as being *at once the minister and interpreter of every emotion and every passion of the human soul.* It is purely subjective; that is to say, all good music is a product of the subjective mind. It is true that the objective mind is capable of directing the muscles so as to play, and play correctly, an intricate piece of music. It is also true that a machine can be constructed upon which any one who can turn a crank can correctly play the same piece of music. The sounds in the two cases are identical in character, — hard, mechanical, soulless. It is only when the muscles have been trained to the point of automatism, to use the common phraseology, or, speaking with scientific precision, it is only when the subjective mind directs the movements of the fingers so perfectly that the objective mind can employ itself with other thoughts, that true music can be produced, the emotions of the musician's soul expressed, or the passions of the listener made to respond. Mechanical music, whether played by a man or a machine, can never inspire the human soul with emotions of love or of patriotism, or lead the warrior to face the cannon's mouth, unless, indeed, the hearer's subjective mind by its interpretive power supplies the missing stimulus. Even the diabolical sounds of a bagpipe, when that instrument is tortured by an inspired native, will move the soul of a Scotchman to a frenzy of patriotism.

To say that the difference consists, not in the instrument that is played, but in the manner in which it is played, would be a trite and commonplace observation. The true explanation possesses a far deeper significance. The difference between good and poor music, or rather between real music and its counterfeit, is determined by the source from which each emanates. Music has its origin in the subjective mind. It is the language of the soul, and is expressive of its every passion and emotion. Like the song of the bird, it is the cry of love, longing, passion, hope. It may also be the wail of despondency and despair, if what the Latin poet tells us be true, —

> "Dulcia defecta modulatur carmina lingua,
> Cantatur cygnus, funeris ipse sui."

In short, it is the most subtle medium provided by Nature whereby soul can speak to soul that language of the emotions which cannot, by a kindred soul, be misinterpreted.

Its primary function is to give expression to those emotions which constitute the motive force in the perpetuation of the species. It is the language of love. Like every other faculty of the soul, it has its normal functions and its abnormal manifestations. It is, necessarily, as foreign to the objective mind as are the emotions to which it gives expression.

But this is a digression. It has been indulged in partly for the reason that I desire, incidentally, to give to music its proper place in the classification of subjective phenomena (for I shall have more to say on the subject hereafter); and partly because it affords a striking illustration of the two-mind theory. Primarily, however, my object in citing the case was to give a fair illustration of the subtle and intricate workings of the law of suggestion, and to point out some of the sources of error against which it is necessary constantly to guard. In concluding this branch of the subject it is only necessary to remark that it seems obvious, even at this stage of the argument, that whilst the two-mind theory certainly affords an ample explanation of all the facts of suggestion, the theory of a single mind is totally inadequate for the purpose, unless we assume two sets of faculties, which amounts, practically, to duality. It is difficult, however, to argue a self-evident proposition; and one is sometimes compelled to draw attention to the absurdities involved in its opposite. In the question under consideration, for instance, the bare presentation of the facts of auto-suggestion compels assent to the two-mind theory as a self-evident solution of the problem. And yet, for the present assuming that a physical demonstration is impossible, it may not be unprofitable to draw the attention of the reader to the absurdities involved in the one-mind theory. To that end let us state the essential proposition relating to the phenomena of auto-suggestion which the one-mind theory necessarily presupposes to be true. It is this: "One mind is able, by auto-suggestion, to convince itself of the truth of a proposition which it knows to be false." It will at once be seen, not only that the proposition involves a palpable absurdity, but it also involves a positive contradiction in terms. Now, there is no rule of logic more manifestly and self-evidently valid than that a proposition involving a positive contradiction in terms is necessarily false. And yet, if we are to accept the one-mind theory as the true one, we must be prepared to accept as true a proposition which, by its very terms, is absolutely untrue. Nevertheless, in one sense of the word, it is true that an *individual* (mark the distinction) may, by auto-suggestion, convince himself of the truth of a proposition which he knows to be false; but that is an absurdly loose and unscientific way of stating the proposition. Stated as follows, it is manifestly true: "The objective mind of an individual may, by auto-suggestion, convince his subjective mind that a proposition is true, which proposition his objective mind knows to be false." The proposition, thus stated, will receive the instant assent of every alienist who has intelligently studied the facts of experimental psychology. He will at least agree that, considered as a working hypothesis for the systematic study of the problems of insanity, the dual-mind theory is perfect. I submit that a perfect working hypothesis is necessarily a true one.

The next proposition in the order of statement is that "the subjective mind is incapable of independent reasoning by the processes of induction." This, as before stated, is a corollary of the law of suggestion; and much of what has been said will apply with equal force to this proposition. It is obvious that the dual hypothesis affords the only solution of the problem; for it is manifestly unthinkable that one mind can at once be capable and incapable of inductive reasoning.

The third proposition, which relates to the power of the subjective mind to reason deductively from given premises to correct conclusions, may be dismissed with the statement that it is not set down as one of the powers which differentiate the two minds by antithesis. The difference is only in degree; but it is so enormous that it must be held to be cumulative evidence of duality.

The same remarks apply with even greater force to the fourth proposition, which relates to the perfect memory of the subjective mind. Its prodigious positive power in that direction, when compared with the feeble efforts at recollection of the objective mind, has all the effect of contrast, and must be considered as an important factor in the problem of duality. It is certainly difficult to imagine one mind as being possessed of two sets, as it were, of faculties, with identical functions differing only in degree, whilst the more perfect of the two is observable only under abnormal conditions of the body.

The fifth and sixth propositions, relating, respectively, to the emotional nature of the subjective mind, and to its power of moving ponderable objects without physical contact, have been sufficiently discussed already. The seventh and eighth propositions, relating, respectively, to telepathy and the abnormality of subjective activity, can be more appropriately discussed in forthcoming chapters. In the mean time it can be truthfully said that the distinctive characteristics embraced in the four propositions last named, present in themselves indubitable evidence of the truth of the dual hypothesis, in that there is no other rational way of accounting for all the varied phenomena which they represent.

The ninth proposition will be discussed in a separate chapter.

CHAPTER XV.

DUALITY DEMONSTRATED BY ANATOMY.

The Brain not the Sole Organ of the Mind. — Surgeon-General Hammond's Researches and Experiments. — The Instinctive Faculties. — The Subjective Mind acts independently of the Brain. — Instinctive Acts Performed after the Brain was totally eliminated. — Children Born without a Brain perform all the Instinctive Functions. — The Medulla Oblongata and the Spinal Cord the Organs of the Subjective Mind. — Idiots without a Brain evince Talent for Music, Mathematics, etc.

THUS far the proofs adduced in support of the dual hypothesis have been confined to the facts of experimental hypnotism and the various other forms of psychic phenomena. This has been done for the reason that in themselves those facts are amply demonstrative of the truth of the hypothesis. But it has often been asked if the facts of cerebral anatomy, physiology, or experimental surgery throw any light whatever upon the subject. This is a pertinent question, because, if those facts are irreconcilable with the hypothesis, the latter must fail under the inexorable rule that one clearly demonstrated adverse fact is sufficient to disprove the most plausible hypothesis. If, therefore, the dual hypothesis is the true one, all the facts of Nature, whether of psychic phenomena or of physical structure, must conspire to demonstrate it. At least, there must be no fact that will disprove it. Thus, if it could be clearly demonstrated that the brain is the sole organ of the mind, the hypothesis of duality must fail for want of a plurality of organs through which a second mind could manifest itself. It is true that the brain itself is dual in a purely physical sense, — that is, there are two hemispheres; but it is demonstrable that they are duplicate organs of the same mind. There is no evidence (except in novels[1]) that the two hemispheres are not identical in function and normally synchronous in action.

If, therefore, duality of mind is to be demonstrated by reference to the physical structure of the animal man, we must expect to find an organ for one of the minds outside of the brain and measurably independent of its conditions or even of its existence. It must, moreover, be the organ of the subjective mind; for it is demonstrable that the brain is the organ of the objective faculties. The organ of the subjective mind must, therefore, be the organ of the instinctive faculties. If it is the organ of the instinctive faculties, it is necessarily the organ of the faculties of intuition and all the others which have been designated as subjective.

Fortunately we have not far to look for demonstrative evidence that the required organ exists, not only in man, but in the lower animals as well. In support of this declaration I shall now cite some passages from the writings of one of the ablest living scientists; namely, Surgeon-General Hammond. What he has said on the subject was written without reference to the dual hypothesis, and certainly without reference to its bearings upon the question of a future life. It has, therefore, all the greater evidential value, for that it was written solely in the interests of pure science, and by one whose professional reputation as an alienist is international, whose works have been translated into every modern language, and are used as text-books by the medical profession in every civilized country.

More than twenty years ago Dr. Hammond delivered an address before the New York Neurological Society, entitled

[1] See "The Hoosier Schoolmaster."

"The Brain not the Sole Organ of the Mind," in which he demonstrated his thesis by a collection of authorities, by original experiments, and by arguments which have never been successfully controverted. In one of his later works[1] he has incorporated the gist of that address in a chapter entitled "The Seat of Instinct." It is from the latter work that I make the following extracts: —

"The brain of man is more highly developed than that of any other animal; he has reasoning powers in excess of those possessed by any living being; his mind governs the world, and, not content with that, seeks for knowledge of those spheres beyond that in which he dwells. But, with all this, he is surpassed by almost every other animal in the ability to perform acts instinctively, — by beings, in fact, whose brains are infinitely less perfect than his, and by others which have no organs corsponding to a brain.

"If the instinct of man were seated in his brain, he would doubtless exhibit a development of this faculty so great as to place him on that score as high as he now stands as regards his mind.

"Going back, for the present, to some of the lower animals, we find that we are able, by certain experimental procedures, to settle some points relative to the seat of instinct with absolute certainty.

"1. *It does not reside exclusively in the brain.* The brain of many animals, especially of those belonging to the class of reptiles, can be removed without the animal suffering any very considerable immediate inconvenience. In such cases the instinct remains unimpaired. Thus Maine de Biran states that, according to Perrault, a viper, the head of which had been cut off, moved without deviation to its hole in the wall. It is impossible that the viper could have seen, heard, smelt, tasted, or felt the wall. It could only have gone toward it instinctively, through the action of a force not residing in its brain, and altogether independent of perception.

"It is an instinct in certain animals to swim when placed in water. I removed the entire brain of a frog, and, after waiting a few minutes for the animal to recover from the shock of the operation, I placed it in a tub of water. It immediately began

[1] A Treatise on Insanity: D. Appleton & Co., 1883. Ch. ii.

to swim. I held my hand so that the animal's head would come in contact with it, and thus further progress be prevented. Continued efforts to swim were made for a few seconds, and then ceased. Removing my hand, the animal again swam.

. . . .

"I have repeatedly performed similar experiments with turtles of various kinds, and lately with water-snakes. In all these cases the whole brain was removed from the cranium, yet the animals did not wobble about aimlessly in the water, but swam straight out into the stream or pond, apparently with as complete a purpose to escape as though they still possessed the full degree of consciousness of the unmutilated animals.

"Such experiments show, beyond a doubt, that perception and volition are not seated exclusively in the brain, and thus that instinct is not indissolubly connected with that organ.

"It is impossible to make similar investigations in the higher animals with such definite results as those obtained with reptiles, but we may call to mind the fact familiar to all physiologists, and to which reference has been made in an earlier part of this work, of the behavior of a pigeon the brain of which had been removed. Though in such a case most of the actions are the result of perception, yet some, as for instance the act of flying when it is thrown into the air, are purely instinctive. But Nature has performed many experiments for us, and these not only on the lower animals, but also on man, which teach us conclusively that even in him instinct does not reside in brain. *They show, too, that certain faculties of the mind are not confined to that organ;* but with that fact we need not at present concern ourselves. [The italics are mine.]

"In certain monsters born without a brain, or with important parts of this organ absent, we have interesting examples of the persistence of instinct. Syme describes one of these beings which lived for six months. Though very feeble, it had the faculty of sucking, and the several functions of the body appeared to be well performed. Its eyes clearly perceived the light, and during the night it cried if the candle was allowed to go out. After death the cranium was opened, and there was found to be an entire absence of the cerebrum, the place of which was occupied by a quantity of serous fluid contained in the arachnoid. The cerebellum and pons Varolii were present.

. . . .

"Ollivier d'Angers describes a monster of the female sex which lived twenty hours. It cried, and could suck and swallow. There was no brain, but the spinal cord and medulla oblongata were well developed.

"Saviard relates the particulars of a case in which there were no cerebrum, cerebellum, or any other intra-cranial ganglion. The spinal cord began as a little red tumor on a level with the foramen magnum. Yet this being opened and shut its eyes, cried, sucked, and even ate broth. It lived four days. Some of these movements were reflex, but others were clearly instinctive, and adapted to the preservation of life.

"Dubois, on the authority of Professor Lallemand, of Montpellier, cites the case of a fetus, born at full term, in which the cerebrum and cerebellum were entirely absent. There were no ganglionic bodies within the cranium, but the medulla oblongata and the pons Varolii. This fetus lived three days; during all this time it uttered cries, exercised suction movements when anything was put into its mouth, and moved the limbs. It was nourished with milk and sweetened water, for no nurse would give it her breast. Dubois cites another case, on the authority of Spessa of Treviso, of a child born without cerebrum, cerebellum, or medulla oblongata, and which lived eleven hours. It cried, breathed, and moved its limbs, but it did not suck. It is difficult to say of this case to what extent its movements were instinctive, and to what extent reflex.

"But all these instances, as well as the experiments referred to as having been performed on lower animals, show that instinct does not reside in the brain.

"2. *It is seated exclusively in the medulla oblongata, or in the spinal cord, or in both these organs.* The observations made and experiments cited under the immediately preceding head, apparently lead to the conclusion that the medulla oblongata, or spinal cord, or both the organs, may be the seat of instinct; and further inquiry shows that this view is as correct as that which associates the brain with the mind."

Dr. Hammond then goes on at length to cite many intensely interesting experiments of his own, demonstrating the marvellous strength and persistency of instinctive acts and emotions after all the intra-cranial ganglia were completely removed. He closes the chapter as follows: —

"In microcephali and other human idiots the instincts are sometimes exceedingly strong, and remain so through life. I have already referred to the instance of one of these creatures, an adult woman, holding a rag-baby in her arms as though it were a child, and in whom the maternal instinct must have been strong, and entirely uncontrolled by the intellect. Some idiots also evince a great instinctive talent for music, and for arithmetical calculations, which, although capable of development, as are other instincts, are nevertheless innate.

"From these facts, and many others which might be adduced in a work specially directed to the consideration of the many interesting points involved, I think it may be concluded that instinct has at least its chief, if not its only, seat in the medulla oblongata and spinal cord. It is possible that the cerebrum, the cerebellum, and the pons Varolii have some influence in strengthening the faculty; but this is not essential, and its exercise is not a mental operation."

It will now be seen that the hypothesis of a dual mind is sustained not alone by the phenomena evoked by experimental hypnotism, but by the physical structure of man himself, as well as of the whole animal creation. It will also be observed that whether we assume two distinct minds, or one mind with two distinct sets of faculties, the logical results are the same. Moreover, if there is but one mind, it is dual nevertheless. It is dual in its organism; it is dual in its functions; it is dual in its faculties; and all the logical conclusions derivable from the hypothesis of two minds are also derivable from the hypotheses of one mind with a dual organism. There is, indeed, one argument for continued existence in the single-mind theory which applies with somewhat diminished force to the theory of duality; for it may be plausibly argued that, if the mind can survive the destruction of one of its organs, there is no good reason why it may not survive the demolition of the other, together with the whole physical structure of which those organs are a necessary constituent element. But this leads directly into the forbidden field of speculation without facts. If a future life is to be demonstrated at all, it will not be by reference to physical structure, but by reference to psychic facts, — the observable phenomena of the soul. When we have learned the fact that there is a physical structure adapted to a dual mental organization, we have learned about the only physical fact, that can possibly be verified, that throws any light upon the subject of a future life or of the relations which the two sets of faculties sustain towards each other. It would be pleasant pastime to speculate upon the possible functions of the medulla oblongata or the cerebellum; but we can have no means of verifying such speculations. The very name and structure, however, of the pons Varolii suggests its possible function as a bridge connecting the domains of the two intelligences, enabling them to hold communion and to act in synchronism; but, in the absence of a possibility of verifying such speculations, we should waste valuable time that could be much better employed in solving the problems presented by the psychic facts which are within easy range of our observation.

I now wish to invite particular attention to two incidental remarks made by Dr. Hammond in the foregoing extracts: "They show, too, that *certain faculties of the mind are not confined to that organ* [the brain.] Again he remarks that "some idiots evince a great instinctive talent for music and for arithmetical calculations, which, although capable of development, as are other instincts, are nevertheless innate."

The importance and far-reaching significance of the remark, and the fact embraced in it, cannot be overestimated. It is a clear and distinct recognition, by one of the ablest living scientists, of the fact that the higher functions of intellect may be, and are by man, performed instinctively, — that is to say, by that mind which operates when the objective senses are inhibited, as in sleep or somnambulism; that mind which is often active and potent when there is a total absence of power in the objective mind, as in idiocy; that mind which performs its functions with undiminished power after the organ of the objective mind has been totally eliminated from the cranial cavity.

The intelligent reader will at once recognize the fact that the musical and mathematical talents possessed by some idiots are illustrations of that class of phenomena which I have elsewhere designated as "the power of intuitive perception of the laws of Nature;" for it is obvious that the musical power of an idiot must be the result of an intuitive or instinctive perception of the laws of harmony of sounds. The idiot's objective mind being extinct, he obviously has no facilities for learning those laws by means of objective education. The same remark applies to the power of an idiot to solve mathematical problems.

Of course, these examples afford only brief and fleeting glimpses into the domain of subjective mental activity and power; but they are all-sufficient to enable us definitely to locate the source and classify the phenomena.

Moreover, it gives us the logical right to infer that faculties which exhibit such prodigious power under abnormal conditions, faculties which are demonstrably not of the brain, have a normal function to perform somewhere; and as they have no normal function in this life, it must be in a life to come. The fact that their greatest observable power is manifested when the functions of the brain are inhibited, gives us the logical right to infer that the physical frame limits its power; and that when it is freed from material limitations, emancipated from the trammels of the flesh, its power of intuitional perception of all Truth will be perfected. Not that its possessor will instantaneously become omniscient. That is a supposition which none but a Hindu philosopher, filled to saturation with that monstrous egotism which results from self-hypnotization, is capable of seriously advancing. The most that can be rationally postulated is that the finite, human entity, endowed with such powers, is invested with the potentiality of rapid intellectual and moral development and indefinite progress.

It will now be conceded that whilst the existence of a dual mind in man is presumptively proven by the very nature of the phenomena exhibited, it is conclusively demonstrated by the facts of physiology and cerebral anatomy. Especially is this true since the facts of cerebral anatomy demonstrate the proposition that there can be no objective mind in the absence of a brain. In other words, as I have before remarked, the objective mind is the function of the brain, and ceases when the brain dies or is destroyed. The subjective mind, on the other hand, belongs to an entity which is neither dependent for its existence, nor for the power to perform its functions, upon the vitality or even the existence of the brain.

It may be asked at this point, "If the objective mind is the function of the brain and they perish together, why may it not be true that the subjective mind is the function of the spinal cord and that it perishes with that organ?" To this I reply that I allude to that question at this point merely for the purpose of showing that I am not unmindful of its pertinency; and I promise to make a full and complete answer to it in the ensuing chapters. The answer to that question pertains wholly to the problem of a future life; and the reader will bear in mind that in this chapter I am merely trying to demonstrate that the human entity has a dual mind; and having done that, I shall reserve the right to hold that, inasmuch as one of those minds clearly belongs to the body, the other as clearly belongs to the soul. It will then be in order to appeal to the facts — the observable phenomena — of the soul to demonstrate its continued existence after the death of the body; and that of itself will be found to be a clear and conclusive answer to the possible supposition that the subjective mind is merely the function of the spinal cord, or of the medulla oblongata, or of the pons Varolii, or of the cerebellum, or of any other organ of the body. That they, or some of them, together with the nervous system, are the *organs* of the subjective entity, by and through which it exercises its control over the functions, sensations, and conditions of the body during its sojourn in it, is quite another proposition, and it may be admitted to be true without argument. It has, however, nothing to do with the question under immediate consideration.

Now let us summarize the foregoing demonstrable and demonstrated facts, so that we may have a clearer view of the effect of the whole. To put them in orderly form, we have the following clearly established propositions: —

1. The objective mind is the function of the physical brain; it is wholly dependent upon the condition of that organ, and it ceases to exist when the brain loses its vitality.

2. Instinctive acts are performed by animals after the brain has been wholly excised from the cranial cavity, and by human beings born without a brain or any other intracranial ganglion.

3. Intellectual feats of a high order, such as playing musical instruments, making mathematical calculations, and many others, belong to the realm of instinct, and are often performed by idiots, — that is, by those destitute of objective intellect.

It must be remembered that these propositions are based upon the facts which have been observed, tested, and recorded by some of the most eminent scientists, living or dead. They are demonstrative of the dual-mind hypothesis; for they show that when one mind is wholly extinguished, either by natural causes as in idiocy, or by a surgical operation such as those described by Dr. Hammond, there is still a mind existent and capable of manifesting itself.

It is thought, therefore, that it may now be fairly claimed that the two-mind theory has been demonstrated to be true: 1. logically, because (*a*) all the facts can be accounted for upon that hypothesis, (*b*) because there is no other hypothesis that can account for all the facts; 2. because the facts of cerebral anatomy, physiology, and experimental surgery all conspire to demonstrate its truth.

CHAPTER XVI.

DUALITY DEMONSTRATED BY EVOLUTION.

Duality in the Lower Animals. — A Primordial Fact. — A Physical Basis for Immortality. — The Ultimate Goal of Psychic Evolution. — Evidence of Design in Psychic Development. — Definition of "Design." — Nature conceals God. — Man reveals God. — The Functions of the Soul. — Design evinced in the Facts of Organic Evolution. — The Benevolence of God. — Painless Death. — The Universal Anæsthetic. — God is ever kind to the Victim of the Inevitable. — Man Re-enthroned.

HAVING now definitely ascertained that the facts of cerebral anatomy not only sustain the dual hypothesis, but locate the organ by and through which the subjective mind can manifest itself, it remains to consider briefly the bearing of the facts of Evolution upon the subject-matter. For, if it is true that man has a soul, and that his soul is immortal, all the pertinent phenomena of Nature will conspire to verify those facts. It will also be in order briefly to consider the facts of evolution, as manifested in the creation of the dual mental organism, in their bearings on the question of the character and attributes of the Deity.

Evolutionists tell us that the tendency of organic Nature from the beginning has been toward the creation of Man. All the facts of physical Nature conspire to demonstrate the truth of that proposition; for, on every line of the evolution of the lower animals upward, the trend of each successive change in physical structure has been toward that final goal. But evolutionists have failed to observe another cognate fact of even more profound significance; namely, that, *in the very beginning of psychic evolution the foundation was laid for the development of an immortal soul.* This foundation was laid when the first organic being, having a dual mental organism, was evolved from protoplasm. And on every line of evolution from the lower animals upward, this dual mental organism is a salient, nay, the dominant, mental and physical characteristic. As I have remarked in a previous chapter, without the dual mind there would be no mental organism remaining after the dissolution of the body and brain, and the consequent extinction of the objective mind. In other words, without the dual mind there would be no basis upon which to build a human soul. With it, existent from the beginning in the lower animals, the process of evolution of the subjective mind from the status of mere animal instinct up to that of an intelligent, self-existent entity, was as natural, as inevitable, as was the evolution of the physical organism through the gradations of animal life up to the perfect physical man. As the soul is connascent with the body, and as the objective and subjective minds are synchronous in action, so has psychic evolution progressed in perfect unison with organic evolution. As the ultimate goal of organic evolution is the creation of man, so is the ultimate goal of psychic evolution the creation of the immortal soul.

On no other rational hypothesis can the wisdom of the Creator be made manifest to finite comprehension. But on this hypothesis, not only is the existence of the God of Jesus made apparent, but his wisdom is demonstrated. It has often been said that there is no evidence of design in the phenomena of physical Nature ; and it must be admitted that in the ordinary phenomena of the physical universe nothing is manifest but the operation of a blind force, which appears to be inherent in matter, operating under the impulse of an iron necessity and apparently in utter disregard of human life or of mercy or of justice. On the other hand, it may be said that this vast universe itself, with its inherent laws, by and through which such grand and beautiful results are brought about, is an evidence of design.

> "Who," says Beecher,[1] "designed this mighty machine, created matter, gave to it its laws, impressed upon it that tendency which has brought forth the almost infinite results on the globe, and wrought them into a perfect system? Design by wholesale is greater than design by retail."

All of which sounds plausible; but it is simply begging the question. If there were no further evidence than that, science would be justified in rejecting the design argument altogether, which it has done ever since the doctrines of Darwin displaced those of Paley.

It seems to me that certain of the specific facts of evolution itself furnish the most indubitable evidence of design that can be imagined. Paley's watch did not furnish a more striking evidence of design than is afforded by the dealings of God with man.

The first question, however, is, what is the proper definition of the word "design," as it is employed as an argument for the existence of God and his intelligent authorship of the universe? This becomes a very important question when we remember how much of man's valuable time is annually wasted in disputatious argument between persons whose only real differences arise from a wrong interpretation of terms.

Design, in the sense which reaches the merits of the question at issue, *is a definite purpose conceived by an intelligent being with a view to the production of a specific useful result.*

This definition applies with equal force to the works of Nature and the works of man. In Paley's watch there was

[1] Evolution and Religion.

evidence of intelligent design; for a specific useful result was produced, — namely, the creation of an instrument for the measurement of time. But a lunatic might produce the most complicated machine imaginable, and if it had no specific purpose to subserve, if it was incapable of being set in motion or of performing any function, there would be no evidence of intelligent design. On the other hand, if, in the works of Nature, there can be found evidence of an intelligent purpose tending at all times toward the production of a *specific* useful result, design becomes as self-evident in Nature as in art. It is, therefore, to a certain extent, a question of evidence as to whether any particular structure can be said to exhibit design; and, within certain limits, what is conclusive to one mind might not convince another. Thus, the theologian finds evidence, conclusive to him, in every tree, leaf, bud, or flower, as well as in the grand structure of the physical universe, of the existence of a God of intelligence and power, and he also sees conclusive evidence of a commensurate design. To this, however, the materialist is ever ready with his reply that all this is the result, not of design, but of the operation of physical laws which are inherent in matter itself; that these blind forces of Nature may go on throughout all coming time, building new worlds and disintegrating the old, in one eternal round, without evincing either intelligence or design, or aught but an iron necessity resultant from immutable laws inherent in the Cosmos. From a purely scientific standpoint, considering alone the cosmogony of the universe as exhibited in the operation of the great physical forces, the materialist is right. It is here that the words of Jacobi strike one with the force of a revelation: "Nature conceals God; man alone reveals him." And it may be set down as an axiom in spiritual philosophy that in the phenomena of mind alone is to be found any evidence whatever of the existence of an intelligence antecedent to the formation of the planetary system, any evidence of design in the structure of the material universe, or any evidence of the immortality of the soul. But in the phenomena of mind we find all the evidence necessary to demonstrate all three propositions. It is only necessary, however, to find evidence of design in order to establish the first two, for it is axiomatic that design presupposes a designer. What stronger evidence of design can be imagined than the fact that all the operations of Nature tend toward the accomplishment of one single specific object? As I have before remarked, all the facts of organic evolution show that the creation of man was the goal toward which all Nature has tended from the beginning. No evolutionist, materialist though he may be, will deny that proposition. This, in itself, is evidence — of a secondary value, however — of design, because in it we find a specific goal. But the materialist will reply that it is inadequate and inconclusive because there is still no evidence of intelligent design, since man, in common with all organic Nature, is destined to die, and in his seventy years of allotted time there is not an adequate motive for bringing him into existence. *Cui bono?* is his ever-ready question; and it is pertinent and unanswerable from his standpoint. If, therefore, there were no further evidence of design than that afforded by man's physical existence, it must be conceded that the argument fails from inherent weakness. But, as I have already pointed out, there was a foundation laid at the earliest possible period in the history of organic evolution for the development of an immortal soul. The dual mind found in each and every animate creature evolved from each individual monad, furnishes this foundation. It is found in all animate Nature, from the lowest to the highest, and on every line both of divergence from man and of convergence toward him; so that, upon whatever line of development the goal might be reached, the embryo of an immortal soul would be present. It is true that there are creatures destitute of anything like a brain, and therefore without an objective mind; but there is nothing in all animate Nature that is destitute of a subjective mind, rudimentary though it be. In fact, the subjective mind antedates the objective, the latter being the result of organic evolution.

It may be asked why it is necessary to assume that the subjective mind in animals is the embryo of an immortal soul in man, since both minds have functions to perform in physical life. To this the self-evident reply is that Nature never produces an unnecessarily complex or complicated organism; and unless we assume that God was incapable of endowing the subjective mind with the capacity to develop the power of reason sufficiently for the purposes of this life, we must assume that it has functions to perform in some other life. We know that it is susceptible to a far higher intellectual development than is the objective mind. The facts, therefore, that the latter was created with functions pertaining exclusively to this life, and that the former was created with faculties that perform no function in this life, constitute demonstrative evidence that the subjective mind was created, *ab initio*, with special reference to a future life. Design, therefore, is as self-evident in the order and process of the evolution of the lower animals up to man as it is in the structure of a watch.

It may be asked why God does not exhibit some tangible, observable, moral attribute toward his creatures, — some benevolent aspect that can bear an interpretation other than that involved in a blind adherence to primordial law. Why is pain and agony, and death and destruction, through the law of the survival of the fittest, necessary for the development of man? The obvious answer is, that it is only through the death of the unfit that room can be made for the survival of the fit. In a word, it is only through death and destruction that evolution is possible. Without death as a universal factor in the physical world, progress would be impossible. And this applies to man as forcibly as it does to the lower animals. If at any moment death should be abolished as to mankind, progress would come to a practical standstill, and the earth would soon be over-populated. This is one of the facts demonstrative that physical life is not the final goal of man or of his evolution; and since it is demonstrable that there can be no higher species developed than man, if he has any future, it must be in another form of existence. To recur to the first question, Why does not God exhibit some tangible moral attribute toward his creatures? — it is answered that he does exhibit a positive, tangible quality of pure *benevolence* towards all animate Nature in a phenomenon that is of such common experience that the world appears to have overlooked it entirely. That phenomenon consists in the absolute immunity from physical or mental suffering in the hour of inevitable death.[1] This immunity is universal in all animate Nature. Moreover, there is every evidence to show that death is a pleasurable process to all who experience it, from the lowest to the highest organism in Nature. Moreover, there is indubitable evidence demonstrative that *God is ever kind to the victim of the inevitable.* This is true whether the inevitable event assumes the form of death or of a surgical operation;[2] for in the subjective state which spontaneously ensues upon the approach of either event, there is provided a universal anæsthetic, which deprives death of its sting and its terrors; and if the surgeon knows the laws pertaining to the subject, it eases the patient of all pain and suffering.

Could further or more tangible evidence be required to

[1] For a full discussion of this subject, see "The Law of Psychic Phenomena."

[2] See "Hypnotism, a Universal Anæsthetic in Surgery," N. Y. Medical Journal for Dec. 22, 1894.

demonstrate the quality of mercy and benevolence in God towards his creatures?

It must be remembered that this immunity from suffering during the process of dissolution cannot be tortured into the domain of attributes resulting from the laws of heredity, or of natural selection, or of the survival of the fittest. It is obvious that when this phenomenon is exhibited, the subject is far beyond the reach of those laws. Nor is there any law or fact in the domain of evolution that can be invoked to explain it. It is a broad, ultimate fact, standing apart, tangible, monumental, demonstrative of the intelligence, the love, the mercy, the benevolence of the Great First Cause.

Is there no evidence of intelligent design in the phenomena of Nature? Let the facts — the observable, tangible, demonstrable facts — answer the question.

Thus is man rehabilitated and re-enthroned as the grand central figure of creation, the ultimate object of creative energy. The Copernican system of astronomy shocked the Christian world, and moved it to deeds of violence by removing the habitation of man from the centre of the universe, and sending it whirling through space, a unit in the Titanic procession around the central sun, — the source of light and warmth and energy. Another shock was sent through Christendom when it was first shown that the facts of evolution proved that God created man, not by miracle, but through the operations of natural law. But Christendom has survived both shocks, and has lived to recognize the fact that science robs not God of his glory nor man of his dignity. On the contrary, Copernican astronomy removed man from the centre of the physical universe only to show him that the central sun, with all its stores of physical energy, is but his *domestic servant*, charged with the duty of rendering his home habitable and beautiful. Evolutionary science removed his origin from the domain of superstition, and revealed him to himself as the central figure in the physical universe. Psychic science proclaims his divine pedigree, confirms his kinship to Christ, and verifies his title-deeds to a "home not made with hands, eternal in the heavens."

CHAPTER XVII.

THE DISTINCTIVE FACULTIES OF THE SOUL.

Every Faculty of the Mind has its Use or Function. — Faculties of the Soul which perform no Normal Function in this Life. — The Man and the Brute psychically Differentiated — Ego-Altruism. — The Instinct of Self-Sacrifice. — Conditions precedent to the Attainment of Immortality.

HAVING now logically and scientifically demonstrated the existence in man of a dual mind, it remains to inquire what are the legitimate logical and scientific conclusions to be derived from that fact. In doing so, the first matter to be considered is the question as to what can be taken for granted. As I have pointed out in the earlier chapters of this book, there is, in every process of reasoning, one factor in the series of steps or propositions leading to a logical conclusion, that is always taken for granted. Thus, in reasoning by induction, we collate our facts and from them we reason up to general principles. That is to say, after the observance of a series of phenomena, when we find a constant recurrence of the series in orderly and unvarying sequence, we are enabled to say with certainty that the same phenomena will continue to recur in the same orderly sequence. In other words we have discovered the law which governs the subject-matter, — the principle which underlies it. But in formulating the law we invariably assume, without formally stating it, the most important proposition in the series which gives it validity. That proposition is that *Nature is ever constant.* It is obvious that if a law of Nature operated in one way at one time, and in the opposite way at another, man would be put to permanent intellectual confusion. He could never be certain that he knew anything. In logical reasoning, therefore, our very silence regarding the constancy of Nature evinces, in terms more eloquent than words, our calm, unfaltering trust in the wisdom and benevolence of the Eternal Lawgiver. So universal is this confidence among civilized peoples, that the proposition that Nature's laws are immutable is not only never questioned, but is, by common consent, regarded as one of those self-evident propositions which to argue is a work of supererogation.

In the domain of psychological research there is also a proposition equally important, and unqualifiedly self-evident, which is assumed without question by all who apply the principles of logical reasoning to the solution of psychological problems. That proposition is that *There is no faculty, emotion, or organism of the human mind that has not its function, use, or object.*

In the physical world it has long since passed into a proverb that "Nothing was ever made in vain." This may be true, and it probably is true; but it is far from being self-evident. For instance, it is difficult to discover the use of mosquitoes, or of venomous reptiles, or of those insects which sometimes destroy the vegetation of whole provinces and bring desolation and famine to the doors of the helpless people. Such things not only cause the unreflecting to doubt the wisdom of God, but they forcibly remind us all of the essential truth of the saying of the pious Jacobi, that "Nature conceals God; man reveals him." No truth is more frequently brought home to the scientific student of Nature and of man than this; for the more we study physical Nature the farther God is removed from us, the more we study man the nearer God approaches to us. Thus, the savage finds God in the most common phenomena of Nature; he sees him in the lightning's flash, and in the thunder he hears his voice. Explain the phenomena to him, and his God is no longer enthroned in the storm cloud; and so it is with all who study only the phenomena of physical Nature. The more one learns of the proximate causes of things, the farther away is removed the ultimate cause; and if one confines his attention to the material universe alone, he is prone to the conclusion that if it is ignorance of Nature's laws that gives birth to gods, a knowledge of those laws will surely destroy them.

Not so in the study of man, of whom it has been truly said that there is nothing else on earth that is truly great, and in whom there is nothing truly great but mind. It is, therefore, only by the study of the mind of man that we shall be able to apprehend the existence of God, to realize his presence, to demonstrate his immanence, or to learn anything regarding the origin and destiny of the soul.

We may assume, then, as the starting-point of our argument, the axiom before mentioned; namely, that "There is no faculty, emotion, or organism of the human mind that has not its use or object." This is a self-evident proposition. It is axiomatic, for it is impossible to conceive its opposite. It is "so evident at first sight that no process of reasoning or demonstration can make it plainer" (Webster). When this can be said of a proposition, all legitimate conclusions to be deduced therefrom partake of the character of the axiom; that is, they are as necessarily true as the axiom itself. They are then scientifically demonstrated.

The first fact in the order of consideration to which this axiom may be applied is that man has a dual mind. This is the primary fact which lies at the foundation of the science of the soul, for it is in a sense demonstrative of the fact that man has a soul; that is to say, if man has a soul, its mental organization must necessarily be supposed to be identified with one or the other of his two minds. Now, it has already been shown that the facts of hypnotism and of experimental surgery demonstrate that the objective mind is a function of the brain, and necessarily perishes with that organ. That fact, therefore, excludes the objective mind from consideration as a possible heir to a future life, and clears the way for the consideration of the facts relating to the subjective mind, which facts must of necessity either prove or disprove its claims to immortality.

The first question to be considered, then, is one of use and function; and the first inquiry is, what purpose is subserved by the two minds? In other words, if man has not an immortal soul, what possible object pertaining to this life can be subserved by the two minds that could not as well have been attained by one mind, since it is axiomatic that Nature never produces an unnecessarily complex or complicated organism?

Now, it must be remembered that so far as this life is concerned, the subjective mind has, primarily, but three functions, namely: 1. Self-preservation; 2. Reproduction; 3. The preservation of the offspring. These may be reduced in terms to one; namely, the perpetuation of the race or species. It has other powers, as we have already seen, and as we shall see more particularly later on. Some of those powers belong exclusively to the subjective mind, whilst others differ only in degree from corresponding powers belonging to the objective mind. But the point to be considered for the moment is that the only *normal* functions performed by the subjective mind during its sojourn in the body, and its connection with it, all pertain to the perpetuation of the species. These are, primarily, purely animal functions, it is true, and are possessed in common with all animate Nature. They are, nevertheless, its only normal functions. This is demonstrated by the fact that *it can never perform any other function or exer-*

cise any other of its manifold powers, except under the most intensely abnormal conditions of the body. This proposition will be amply demonstrated in a later chapter. For the present it must be taken for granted for the sake of the argument which follows.

Is it conceivable that a power which is capable of creating man and endowing him with powers that we know him to possess, could not have endowed his objective mind with the instinctive powers and functions which have been named? In other words, what is the use of such a complex and complicated organism as is involved in the mechanism of a dual mind when a unitary mind could just as well have performed all the normal functions now performed by both?

It will thus be seen that the bare fact that man possesses a dual mind necessarily leads to the conclusion that one of them does not pertain exclusively to his physical well-being.

Let me not be misunderstood at this point. I do not undertake to say that immortality is directly foreshadowed by the fact that man has a dual mind, and that his subjective mind is endowed with the faculty of animal instinct. But what I do say is that the fact that he has a dual mind, indirectly but inevitably, leads to that conclusion. It is, indeed, the fundamental basis of fact from which we may proceed to a demonstration of a future life. It is a logical as well as a physical necessity, since the objective mind is merely a function of the brain. There must, therefore, be another mind, another intellect, another entity in man, or there would be nothing upon which to predicate the hypothesis of immortality. It is, therefore, an argument for a future life in that the existence of two minds cannot be accounted for on any other rational hypothesis.

I cannot be unaware that the argument is far from demonstrative so long as none of the attributes and powers of the subjective mind have thus far been taken into consideration except the purely animal instincts which we have been considering, since those functions pertain wholly to the physical plane. If all else were left out of consideration, man would be on an exact level with the brute creation, so far as the question of a future life is concerned. It would be impossible to postulate a rational hypothesis of immortality for man that would not apply with equal force to the brute, if the soul of neither was endowed with higher attributes than the instincts which belong alone to the physical plane.

It is just here, however, that the difference between man and brute becomes manifest; and it is upon these lines that must be settled the vexed question whether the brute, equally with man, may not be able to read his title clear to mansions in the skies, since the former, in common with the latter, is endowed with both objective and subjective faculties.

The subjective faculties of the brute are limited to those primary instincts which pertain wholly to the perpetuation of the species, and belong therefore exclusively to the physical plane. On the other hand, the subjective mind of man is endowed with intellectual faculties which far transcend those of the objective intellect, and some of which perform no normal function on the physical plane.

Nor must it be forgotten in this connection that the instinct of self-preservation itself, common as it is to all animate Nature, and purely selfish as it is among the brute creation, contains within itself the elements of the purest altruism. And when it is normally developed in man, it becomes the most purely unselfish of all the human instincts, and exhibits itself in the noblest acts of self-sacrifice recorded in history, the sublimest heroism conceivable by the human imagination. It was regnant in the devoted band who held the pass at Thermopylæ; in "Horatius, who held the bridge in the brave days of yore;" in Winkelried, who swept the enemy's spears into his own body, that he might break the Austrian phalanx; in the pilot who, with arms shrivelling in the flames, guided his burning ship toward the shore, to the end that others might survive; in the captain of the sinking ship who stood at his post until the last passenger was safe, and was alone drawn into the vortex; in him who yielded up his life on the cross in testimony of his divine mission to "bring life and immortality to light,"—in all those noble, self-sacrificing souls who suffer and die that others may live. These are the higher aspects of the instinct of self-preservation; for that instinct in man pertains not alone to the preservation of the individual life, but to that of the species to which it belongs. It is just as much a matter of instinct to sacrifice one's own life for the preservation of the lives of others as it is to shrink from imminent peril when one's own safety is alone involved. In a word, when the lives of others are at stake, cowardice is a purely abnormal manifestation of that instinct. Heroic self-sacrifice, when others are in peril, is alone normal. It will thus be seen that the principle of "the greatest good to the greatest number" is the ultimate form of the evolutionary development of what has been regarded as the most purely selfish instinct. In the lower animals it is a purely selfish, individualized instinct. In man it rises to the dignity of what, for the want of a better term, may be designated as *ego-altruism*,—the *ægis* of all humanity.

It may be asked, "Why, if every faculty has its use or object, is the brute endowed with two minds, if neither of its minds is destined to a future life? In other words, why does not the dual-mind hypothesis argue immortality for the brute as cogently as it does for the man?" It may as well be asked, "Why does not the possession of two eyes prove that the brute is destined to become a man?" Or, "Why does not the presence, in various species of animals, of the rudimentary physical structure of man, argue that each individual animal is destined to become a man?" The answer to all these questions is the same; namely, *The creation of the physical and the psychical man is the goal toward which Nature has tended from the beginning; but it was through the processes of evolution that both body and soul were created.*

Man's physical structure, in rudimentary form, is found in the animals from which he was evolved; but it does not follow that the individual animal is to become a man. Man's psychical organism is found, in rudimentary form, in the brute creation, the same dual mental organism being present in all animate Nature; but it does not follow that the individual brute is to inherit immortality. The rudimentary form of man in the animal rendered it possible for the processes of evolution to culminate in the creation of the perfected physical man. The rudimentary psychical organism in the animal rendered it possible for evolution to elevate the embryotic soul to the full stature of a living, conscious, individualized entity, capable, under certain conditions, existent in man alone, of sustaining an existence independently of the physical organism.

The primary condition precedent to the attainment of such an existence is necessarily that of consciousness. It is axiomatic that no individualized existence worthy of the name can be sustained by a living organism, physical or psychical, in the absence of consciousness. It is also obvious that an animal can have no consciousness of the possession of a soul. Nor can the soul be conscious of itself in the absence of any suggestion or information conveyed to it by objective education. Jesus, who was master of the science of the soul, drew the line, on strictly scientific principles, between the man and the brute, when he proclaimed the law that belief — faith — was the essential prerequisite to the attainment of immortal life.[1] "Faith," in the sense in which Jesus employed the term, means much more than "belief," although the latter is included in the

[1] For a fuller discussion of this important problem, see "The Law of Psychic Phenomena," ch. xxv.

term. Faith, in the psychic sense, and that is the sense in which Jesus employed it, is *conscious potentiality*. It is a power; it is *the* power of the soul. All psychic phenomena demonstrate that proposition. Without it there can be no psychic phenomena beyond the exercise of the purely animal instincts. It is the creature of suggestion. Suggestion alone awakens it into existence; suggestion can utterly destroy it. Inasmuch as no suggestion of the possibility of immortal life can be conveyed to the embryotic soul of the brute, the conscious potentiality requisite to the sustentation of independent existence does not exist; and it obviously cannot exist in other than an intelligent being. And this remark, according to the philosophy of Jesus, applies to all the brute creation, whether it is embodied in the form of animals or of men.

It is thought that enough has now been said to make a *prima facie* case, because I have shown: 1. That there is a basis in the mental organism of man upon which the hypothesis of a future life can be postulated, in that (*a*) there are two minds, (*b*) the subjective mind does not necessarily perish with the brain; 2. That no other rational hypothesis, which will account for all the facts, psychical and physical, has yet been formulated.

CHAPTER XVIII.

FACULTIES BELONGING TO A FUTURE LIFE.

The Necessity for limiting the Powers of the Subjective Mind in this Life. — Man a Free Moral Agent. — The Law of Suggestion a Necessity. — Limitations of Power pertain only to this Life. — Induction unnecessary in the Future Life. — Intuition takes its Place. — Induction Impossible when the Power of Perception exists. — The Higher Intuitional Powers Useless in this Life. — The Power of Correct Deduction in Man and Animals.

WE will now proceed to the consideration of those peculiar powers, functions, and limitations of the subjective mind which seem to be especially adapted to a future life. In doing so, the mental faculties will, for convenience, be divided into three classes, namely: 1. Those which belong exclusively to the subjective mind; 2. Those which belong exclusively to the objective mind; and, 3. Those powers possessed in common by the two minds, differing only in degree. They will be considered as nearly as possible in the order named, although it will be necessary in some cases to group two or more and consider them together.

Before proceeding, however, I desire to impress upon the mind of the reader the fundamental axiom mentioned in the chapter preceding; namely, that *There is no faculty, emotion, or organism of the human mind that has not its function, use, or object.* A moment's reflection will be sufficient to extort the assent of every logical mind to this proposition. If any one will try to imagine the contrary or opposite proposition to be true, he will find that it is absolutely and unqualifiedly unthinkable. I lay particular stress upon this proposition for reasons that will more fully appear as we proceed. In the mean time it will be obvious to every logician that any legitimate conclusion derivable from a proposition so far-reaching and so perfectly self-evident, must necessarily be invested with a profound significance.

We have now three fundamental propositions to start with, each of which is either self-evident or is demonstrable by reference to the facts of experimental psychology, cerebral anatomy, or experimental surgery. They are: —

1. Man has a dual mind.

2. Each of the two minds has powers, functions, and limitations which clearly differentiate it from the other.

3. Each power, function, and limitation necessarily has its use, function or object.

The first and second of these propositions have been clearly demonstrated by the facts of experimental psychology, cerebral anatomy, and experimental surgery. The third is axiomatic.

I will now add a fourth proposition which will complete the chain of logical premises necessary to a complete demonstration of a future life for mankind. It is this: —

4. There is no power, faculty, function, or limitation of the subjective mind, which is peculiar to itself and which clearly differentiates it from the objective mind, that has any normal use or function in a purely physical existence.

No one will deny that, if this proposition can be substantiated, the conclusion that man is heir to a future life is irresistible; for if every faculty has its use, and the subjective mind has faculties that are of no use in a physical life, it follows that those faculties pertain to a life or existence untrammelled by physical limitations. This conclusion is as scientifically correct as it would be to predicate the capacity to navigate the air, of an animal with wings; or the capacity to fly, swim, and walk, of a fowl with wings and webbed feet.

If anything were needed to add logical weight to this argument, it will be found in the fact that the converse of the fourth proposition is also true; namely, —

There is no power, faculty, function, or limitation of the objective mind, which is peculiar to itself and which clearly differentiates it from the subjective mind, that could have any possible use or function in any but a physical life.

Standing by itself, this proposition is purely negative so far as the question of a future life is concerned. But, considered with reference to the fourth proposition, it affords a contrast which is as striking as it is important, in that it demonstrates that the two minds are adapted to two different planes of existence, and that neither is adapted to the other.

We will first consider the bearing of the law of suggestion, and its corollary, the limitation of the reasoning powers of the subjective mind, together with its power of intuitive perception or cognition of the laws of Nature. As we have already seen, these powers and limitations are peculiar to the subjective mind. It has also been pointed out that the limitation of power incident to its control by suggestion is a necessity, for the obvious reason that one mind must necessarily be normally under the absolute control of the other if harmonious relations are to be maintained.

There are an infinite number of reasons why the objective mind should be invested with that responsibility, and I certainly know of none against it. The first, and perhaps the most important, is that in no other way could the objective man — the human entity — be made and held responsible for the moral status of his own soul. In other words, it is by this means, and by this alone, that man is constituted a free moral agent. The objective man, being endowed with the power to reason by all methods, his mind being pure intellect and destitute of emotion, is manifestly best fitted for the exercise of that judicial power — that absolute sovereignty — which must of necessity reside in one or the other of the two minds. On the other hand the subjective entity, being the seat of the emotions, and charged in this life with but the three normal functions, which constitute the master passions of all animate Nature, must necessarily be under the dominion of some moral force capable of restraining and regulating those passions and directing their current into legitimate channels. Otherwise man could never have been elevated above the level of the brute; certainly not above the status of the most primitive savagery. Civilization would be impossible, morality would be nameless, and religion non-existent. It will thus be seen that for the purposes of this life the law of suggestion is a necessity. If this life were all, and the three normal functions of the subjective mind were all that pertained to it, it might well be asked, why the necessity of a dual mind? And it would be difficult to find a rational answer. But when we postulate a future life for man, we find ample reason, not only for two minds, but for the limitation of power in the subjective mind. For it must now be observed and borne in mind that those limitations of power pertain exclusively to this life. But why the necessity for limiting the reasoning faculties of the subjective mind, thus depriving it of that power which invests the objective mind with its supremacy and dominion over the forces of Nature? There are two answers to this question. The first is that it is a corollary of the law of suggestion; for that law could not exist if the subjective mind possessed the power to reason independently, and it is therefore an absolute necessity for the purposes of this life. The second answer is twofold, —

Firstly, there is obviously no necessity for the subjective mind to possess such a faculty, even if it were possible for the two minds to exist together in harmony, since the objective mind possesses it and is the controlling and responsible power in this life. Secondly, *the power of the subjective mind to reason inductively is neither necessary nor possible, for the reason that it is endowed with the faculty of intuitive perception or cognition of Nature's laws, independently of objective education.*

A few words will make this proposition plain to the most superficial reader. What is induction? Induction is a method of inquiry. It is the slow and laborious process of investigation by which the dull and plodding objective mind of man is enabled to learn something of the laws which govern the universe. It is the one faculty which enables him to be certain that he knows something. It is the one weapon which enables him to conquer the forces of Nature. It lends accuracy to learning, permanency to progress, and stability to civilization. But it is of the earth, earthy; for it belongs alone to the objective mind. It is a function of the brain, a product of organic evolution, a faculty developed in response to the necessities of man's physical environment. I repeat, — it is of the earth, earthy. It can have no place, or power, or function in the future life, for the simple reason that it is not a necessity of that plane of existence. It is not a necessity, for the reason that the soul possesses that power of intuitive perception or cognition of Nature's laws which renders any process of laborious inquiry in the nature of induction superfluous, — impossible. And this is why I have said that inductive reasoning is neither necessary nor possible to the subjective mind. It perceives; and its power of perception as far transcends the power of induction as Omniscience transcends the powers of sense. It is, in fact, the power of Omniscience, and its possession by the human soul demonstrates its kinship to God; for *God himself cannot reason inductively*. Induction, as I have before remarked, is a finite process of inquiry into something which finite man does not already know. To suppose an omniscient God to be capable of induction would be a contradiction in terms as gross as it would be to say that a triangle is rectangular.

Again, it must be remembered that this power of intuitive perception of Nature's laws has no legitimate place in earthly life. This is amply demonstrated by the fact that only occasional glimpses of phenomena can be obtained which render it certain that the power exists as a part of the mental equipment of the subjective mind; and these glimpses can only be obtained under the most intensely abnormal conditions of the body or of the objective mind, or of both. This is a fact within the knowledge of the most superficial observer, but it will be more fully dealt with in a subsequent chapter.

Furthermore, its lack of legitimate function in this life is shown by the fact that, outside of the domains where demonstration by other means is possible, we can never be certain of the verity of subjective perception, owing to the ever-present power of suggestion. That is to say, when the power of perception is exercised, say, in the field of mathematics, we have the means of testing the accuracy of the alleged perceptions; but when we have not the means of verifying the alleged facts, we can never avail ourselves of the alleged information, for the reason that we cannot know to what extent the law of suggestion has operated as a factor in the case.

It comes to this, therefore, that we can never be certain of the accuracy of alleged intuitions, unless they are otherwise verified; and they cannot be verified except by the exercise of the powers of objective reason.

It follows that, so far as the mental operations of this life are concerned, the subjective powers of intuitive perception are superfluous and useless. If they were not so, that is, if they were normal, and could be depended upon as a source of information, the objective powers of reason would be superfluous, and therefore useless.

Before we close this branch of the subject, one word must be said concerning the transcendent power, possessed by the subjective mind, of correct deduction from given or suggested premises. Although this power differs only in degree from the corresponding faculty possessed by the objective mind, it must, for obvious reasons, be considered in connection with the other reasoning faculties. This faculty is an essential concomitant to the law of suggestion. The power of suggestion would be of little avail if the subjective mind could not correctly deduce all legitimate conclusions from the premises embraced in a suggestion. It is true that wrong or absurd suggestions will lead to wrong and absurd conclusions; but the conclusions will be logically correct whether the premises are true or false. This is inevitable from the very perfection of the faculty of deduction; but it is compensated for in many ways, for it becomes a factor of the utmost value when a correct suggestion is made. For instance, in the moral training of the subjective mind of a child, if it is punished for stealing from or lying about John Doe, the lesson that it learns is, not simply that it is wrong to injure John Doe, but that it is wrong to tell a falsehood or to appropriate the property of others. It is, however, too obvious to need illustration, that no suggestion could be *intelligently* carried into effect in the absence of this faculty of logical deduction.

The same faculty is possessed by animals; and, together with the power of suggestion over the animal kingdom, it constitutes the prime factor in the combination of causes which enables man to assert and maintain his dominion over the beasts of the field. A single illustration will suffice. The first step which an intelligent trainer takes in breaking a horse is to throw the animal, and hold it down until it ceases to struggle. When this is accomplished, half the battle is won; and although other means to the same end may be adopted, they all tend to demonstrate to the horse that his trainer has absolute power and dominion over him. The rest is easy when gentle kindness and persuasion are employed to teach the animal his duties. Now, the first and most essential step named constitutes a suggestion to the animal that his trainer possesses complete mastery over him, and that it is useless to struggle against superior physical force. This suggestion, however, *per se*, applies only to the individual trainer; and but for the faculty of deduction, no one but the trainer could drive the animal. But the horse, from the suggestion that his trainer has power over him, deduces the conclusion that other men possess the same power. Otherwise every new driver would be obliged to rebreak the horse.

It will thus be seen that the subjective faculty of correct logical deduction from suggested premises possesses a far-reaching significance, and importance in matters of every-day experience in this life. Concerning the part it may play in the mental operations incident to the life to come, it would, perhaps, be useless to speculate; although its concomitance with the faculty of intuitional perception is too obvious to require comment.

Having briefly discussed the reasoning powers of the two minds, we may now pause to take our bearings and find where we stand at this stage of the argument.

We have located and found a use for every reasoning or intuitional faculty of the two minds save one. We have found: —

1. That the faculty of induction belongs exclusively to the objective mind, and hence pertains exclusively to earthly life.

2. That the faculty of intuitional perception belongs exclusively to the subjective mind.

3. That this faculty of intuitional perception performs no normal function in earthly life, as is clearly shown by reference to the facts, —

a. That we catch only occasional glimpses of that faculty in the subjective mind, and know with certainty of its existence only by and through abnormal means and the most intensely abnormal conditions of the objective mind and of the body.

b. That, owing to the law of suggestion, no conclusions arrived at by alleged intuitional processes can be relied upon in this life unless they are verified by objective methods of investigation.

c. That the labor incident to verification is at least equivalent to that of making an original investigation of the subject-matter.

d. It is, therefore, not only abnormal, but superfluous and worse than useless on the physical plane.

The conclusion seems irresistible that at least the purely intellectual part of the subjective entity belongs exclusively to a future existence.

CHAPTER XIX.

THE DYNAMIC FORCES OF THE MIND.

The Buddhistic Nirvana. — A purely Intellectual Existence without Memory, Emotion, or Personality. — The Basis of their Philosophy. — Incomplete Observation of Psychic Powers. — Ignorance of the Law of Suggestion. — Requisites for the Retention of Personality. — Memory. — Consciousness. — Will. — Will is Desire. — The Strongest Desire of the Soul. — Egoism and Egotism of the Soul. — Egoism the Normal Desire for Retention of Personality. — Egotism Abnormal Self-Conceit. — The Dynamics of the Soul. — The Kinetic Force of the Soul.

IT has now been shown that the subjective entity possesses all the mental equipment necessary for an enjoyable existence as a purely intellectual being, without being the possessor of any of the faculties which have been designated as belonging exclusively to the objective mind.

It may be remarked, in passing, that the possession of the intuitional faculty alone would, to the disembodied soul, constitute practically the Buddhistic Nirvana, and would doubtless nearly approach the ideal of the average Yogi, who begins his search for divine illumination by severing every domestic tie, repudiating every social obligation, suppressing every human emotion, and strangling every human affection; and pursues his quest for truth by sitting on his haunches, thinking about himself, and trying to stare his umbilicus out of countenance. He seeks for "emancipation" from every human passion, and contemplates with calm indifference the prospect of the annihilation of his individuality: he longs for absorption into the Deity, for rest in Nirvana. It is, perhaps, not strange that the average Hindu should regard absolute rest as the acme of human felicity. His climate, his social and political environment, his diet, his habits of body and of thought, the fauna of his native land, and the character of his Western proselytes, — all have a tendency to aggravate that feeling of weariness which seems to involve both body and soul, and to be congenite with the whole Oriental race.

The Hindu philosophy of a future life is based largely, if not wholly, upon an observation of that one faculty of the subjective mind to which I have just alluded; that is to say, the salient feature of the subjective phenomena of the Yogis, *et id genus omne*, consists in entering that hypnotic state known as "ecstasy." In that state they become "illuminated," as they term it; and they imagine that they come into direct communion with the Deity, and that they are put in possession of all knowledge, and a large share of the deific power. In short, they identify themselves with the Deity, in imagination; and they come to the conclusion that they have penetrated the secret of a future life, and are enabled to define its conditions. Now, although there are as many different sects in India as there are in Christendom, and although their views are as widely diverse regarding non-essentials, yet they all agree upon one point; namely, that the ultimate destiny of man is to be absorbed into Deity, and identified with him. In the Buddhistic philosophy this means utter annihilation of individuality. Of course the different sects hold diverse views even on this point; but this seems to be the general trend of both the Brahmanist and the Buddhist doctrine of a future life. That question, however, is unimportant for our present purpose. The significant point is that they have arrived at the general conclusion, from an observation of the phenomenon of ecstasy, that soul is ultimately absorbed into the Deity, and thereby comes into possession of all knowledge, power, and dominion. Now, that condition is identical with the one in which the faculty of intuitive perception or cognition of truth is oftenest observed. It is seen in other states or conditions, to be sure, from that of apparent normality up to ecstasy; but the last condition is the one in which it is most frequently observed, and it is the one in which it can be experimentally reproduced. It is doubtless true that the Indian adepts have occasionally found that some of their intuitional perceptions could be verified, as by some mathematical process; or the faculty of telepathy may have been developed, and the information obtained in that way may have occasionally been found to be veridical. A few circumstances of that character would inevitably lead that class of minds to the conclusion that all impression felt while in the ecstatic condition were revelations of divine truth. I say that that conclusion would be inevitable, just as the phenomena of spiritism has inevitably led the same class of minds to the conclusion that they are produced by spirits of the dead. But this must be said in extenuation of both conclusions; namely, that until the law of suggestion was discovered, there was no other rational hypothesis which could explain all the phenomena. In the mean time there were but two paths open to the scientific mind. One was to accept the phenomena for what they purported to be; the other to deny their existence. To explain them on scientific principles, in the absence of any knowledge of the law of suggestion or of the power of telepathy, was impossible. These discoveries, however, have changed the whole aspect of the questions involved, and relegate the visions of ecstatics to the category of subjective hallucinations induced by suggestion.

If anything were needed to demonstrate the proposition that ecstatic visions do not reveal scientific truth, it would be found in the fact that there are many different sects in the Orient, each of whose doctrines is based upon the phenomena of ecstasy, and yet they are so divergent that they are scarcely recognizable as springing from the same root. Surely, if ecstatic visions reveal divine truth, there can be no room for difference of opinion among those favored mortals to whom it is thus revealed.

The truth is, that the whole of the vast fabric of Oriental philosophy is based upon psychic phenomena produced in utter ignorance of the law of suggestion; and as that law is fundamental, universal, and never-failing in its operations so long as the soul inhabits the body, it follows that Hindu philosophy is destitute of any scientific value whatever.

The ever-ready answer to this is the declaration that in the ecstatic condition the subject is lifted into a higher spiritual realm, where he is exempted from the laws which govern ordinary mortals, where the law of suggestion no longer prevails, and where all truth stands revealed to the "disenthralled cosmic consciousness," whatever that may mean. In other words, they simply beg the question.

It is noticeable that the Oriental ecstatics have many and a constantly growing number of feeble imitators in the Western world, who are assiduously practising self-hypnotization with a view of coming into "conscious communion with the Deity," etc., etc. They, too, are instructed that in the "higher spiritual realm" they are exempted from that inconvenient law of suggestion which prevails so extensively among less favored mortals "on the lower psychic plane," and that consequently their visions are veridical — their imaginings are the essence of divine truth imbibed directly from its divine source. It is useless to remind that class of minds that all the facts of psychic science conspire to demonstrate that the law of suggestion is the universal, ever-present, all-potent factor which governs the manifestations of the soul so long as it retains its connection with the body. It is useless to challenge them to produce one fact that militates against that proposition; for scientifically verified facts are not to their liking when such facts do not harmonize with their emotions. It is a thankless task to warn them that such practices are abnormal to the last degree, destructive to the nervous organization, weakening to the objective intellect, and, instead of being promotive of spiritual growth, constitute the direct path to spiritual imbecility. It is even useless to try to encourage them by explaining to them that the faculties which they are trying to use amidst the trammels of the flesh and the limitations of the law of suggestion, are faculties which are normal only to a future life, when the soul is freed from its earthly limitations, and is thereby enabled to exercise those intuitional powers of perception which belong only to a realm of truth. They are not content to await their allotted time, but rush unbidden to the gates of heaven, determined to penetrate the secrets which Jesus withheld from all mankind, and which must forever remain a mystery to incarnate man.

I remarked, in the beginning of this digression, that "the possession of the intuitional faculty alone would, to the disembodied soul, constitute the Buddhistic Nirvana." I mean by this remark that, if this faculty of the soul constituted the only one which it carries with it into a future life, it would correspond exactly with the Brahmanist idea of the status and capacity of the perfected soul. It would be a purely intellectual being, destitute of emotion, and therefore divested of all human interest or affection, bereft of memory, and therefore of individuality, and possessing only what is vaguely termed a "cosmic consciousness," which seems to have reference to nothing of human interest, if it has any definite meaning whatever. We might well suppose such a being to be destined to be absorbed into Brahm without either loss to itself or material gain to Brahm.

To the Western mind, accustomed to regard a future life

as one fraught with human interest, such a destiny would be regarded as equivalent to utter annihilation. It becomes important, therefore, for us to inquire whether there is anything in the structure of the subjective mind to warrant a conclusion so repugnant to every normal emotion of the human soul.

In pursuing this inquiry, the first question which naturally arises is, What is the primary and most essential attribute or power which man, constituted as he is in this life, most naturally desires his soul to possess as a means of enjoyment in a life to come? Clearly, the answer of every normally constituted person would be, *The retention of conscious individuality*. It is obvious that all other enjoyments depend upon that. Any condition, *minus* the personality, would be the equivalent of annihilation. "What a man is and has in himself," says Schopenhauer, "in a word, personality, with all it entails, is the only immediate and direct factor in his happiness and welfare." This, in the very nature of things, must be as true of a future life as it is of the present. If, therefore, it can be shown that the soul has the means and the power to retain that personality, it will be seen that the great factor in man's happiness and welfare will be present in a future life.

The three essential prerequisites to the retention of personality are (1) Consciousness, (2) Memory, and (3) Will. Consciousness and memory are the two co-ordinate, concomitant factors which constitute the personality of each individual, so far as he himself is able to realize it, or to take cognizance of his own existence as a distinct entity. Consciousness is the state of being aware of one's existence and of one's mental acts and states. Memory is the faculty of the mind by which it retains the knowledge of previous thoughts or events; without memory there could be no possible retention of personality. What is known of the sum total of a man's experiences and qualities constitutes his personality as it is cognized by others. What one remembers of those experiences constitutes his personality as cognized by himself. It follows that one's personality is more or less pronounced in proportion to the retentiveness of his memory; just as one is distinguished in the estimation of his fellows in proportion to what is known of his characteristics as shown by the known events of his life. If, therefore, the memory and the consciousness should be blotted out, the being might sustain an existence, but it would be purely vegetal. The personality would remain only as a memory of those who knew him; but, so far as the individual would be concerned, his condition would be the equivalent of annihilation.

The intelligent reader will have anticipated me in what I am to say regarding the perfect memory of the subjective mind, and the conclusions derivable therefrom. As the perfection of subjective memory has been again and again demonstrated,[1] it will here be taken for granted. The general conclusion to be derived is that the personality of the soul will be as much more pronounced than that of the objective, physical human entity as the memory of the former exceeds that of the latter.

Again the reader must be reminded of the fundamental axiom upon which this argument is based, which is, in brief, that there is no useless faculty of the human mind. It will then be pertinent to inquire what possible function a perfect subjective memory can be supposed to perform if it is not what has been indicated? It must be remembered, —

1. That the objective mind possesses a power of recollection which is all-sufficient for the uses of this life. A perfect subjective memory has, therefore, no function to perform in the intellectual processes of objective existence.

2. That all exercise, on the physical plane, of the powers of the subjective mind are abnormal, and productive of

[1] See "The Law of Psychic Phenomena," ch. iv., v.

untoward results to the physical frame. It has, therefore, no normal function pertaining to physical life.

The question then arises, What function can it perform in a future life? Clearly it is not a necessity as an aid to the intellectual processes of the soul; for that has been shown to possess the inherent power of intuitional perception or cognition of all truth. It would, therefore, be a faculty as superfluous to the soul as the faculty of reasoning inductively, considered solely as an aid to intellectual development, and for the same reason.

There remains but one other direction in which we can look with reasonable hope to find a normal function for perfect subjective memory. In that direction we find three functions, two of which are concomitant and obvious, and the other is inferential and speculative.

The first is what has already been indicated; namely, it enables the soul to retain its personality. The second is included in the first; namely, it enables the soul to recognize its friends, and, inferentially, to resume social relations at will. Here, then, are ample functions for the memory of the soul, and they are of weighty import in more senses than have been named; but these will be reserved for future consideration. It is sufficient for the present to have shown that the soul has the prerequisite faculties for the retention of its personality, and it remains to ascertain if it also possesses the motive force necessary for the purpose. The third prerequisite named is *will*.

Will is a motive force: it chooses when stimulated by desire; it has its biological origin in desire. One's will, therefore, is strong in proportion to his desires regarding any particular object; and, other things being equal, his ability to accomplish a particular end is in exact proportion to his strength of will, or desire to do so. Without will, therefore, nothing can be accomplished. It is this which distinguishes the man from the brute; that is to say, it is one of the factors which count for immortal life in man, but which is totally absent in the brute; the man wills to retain his individuality after the death of the body, and he alone has the power and potentiality of a self-existent entity. The inchoate soul of the brute has no conception of a future life, and hence no desire — no will — to enable him to retain his individuality, and no consciousness of the possibility of any but a physical life.

The strongest desire of the human soul is to retain its personality. It is instinctive; that is to say, it is the extension of the instinct of self-preservation to a future life. It is the higher manifestation of that instinct, the primary function of which is the preservation of the body, but which belongs equally to the soul, and performs its higher function in the preservation of its personality. Jesus expressed, in the strongest possible terms, the strength and intensity of human desire, when he said, "What shall it profit a man if he shall gain the whole world, and lose his own soul?"

If any one doubts the intensity of the desire of the normal man for the retention of his personality, let him try to think of some living person with whom he would be willing to exchange personalities. It is doubtful if there is a person so utterly miserable and unfortunate that he would be willing to blot out all of the memories which constitute his individuality, abandon all his hopes for the future, and accept in exchange the personality of the most fortunate person on earth. Such is the inherent *egoism* (not egotism) of the subjective entity.

Let not this word be construed altogether in the offensive sense; for the emotion represented has a normal function to perform which is of the most transcendent importance.

It is desirable at this point to understand clearly certain very important distinctions which must be drawn before the subject can be properly understood. As the dictionaries do not make the proper distinctions, *eo nomine,* it will be necessary to depart from the usual definitions. The word "egotism" being now always employed in the offensive sense, extreme vanity and self-conceit being designated by that term, it will be retained to designate the abnormal manifestation of the attribute to which I wish to draw attention. The word "egoism" has always been defined as synonymous with "egotism," except where it is employed in reference to a certain school of philosophy supposed to have been founded by a sect of the disciples of Descartes and Fichte. But I shall press the word "egoism" into service to designate the normal manifestation of the attribute or faculty alluded to, partly to avoid coining a word, but principally because it happily expresses my exact meaning. I define "egoism," therefore, as *the passionate desire of the human soul for the retention of its personality or individuality.* This represents the normal emotion and its normal manifestation; and, as before remarked, it is the strongest emotion of the human soul. Its inherent strength and persistency is made apparent to common observation in the fact that it manifests itself under all conditions of abnormal subjective mental activity. Under abnormal conditions, that is to say, when for any cause the subjective mind is in control, it is the one never-failing characteristic emotion, dominant over every other, and as offensive in its manifestations as it is colossal in its proportions. It is ever present in habitual hypnotic subjects, especially in those who are used for stage exhibitions; it is chronic in adepts, spirit mediums, and criminals; it assumes colossal proportions in actors, artists, poets, and musicians; and it is the salient characteristic of hysteria, genius, and insanity. Nor is it confined to the classes named; for it characterizes all who for any cause allow the subjective faculties to attain undue prominence in earthly life. These are its abnormal manifestations, and they are abnormal simply because they are manifested under abnormal mental and physical conditions. This is "egotism" in its offensive sense; and it is the primary, fundamental emotion which, uncontrolled by objective reason, leads to immorality, vice, crime, and insanity. It belongs exclusively to the subjective mind.

The corresponding attribute of the objective mind can be appropriately designated by no other term than that of "self-respect;" which is concomitant with reason, and springs from a consciousness of rectitude and of power equal to all legitimate demands. It is that regard for one's character which manifests itself in abstention from the performance of unworthy actions. It is always normal when controlled by reason; but when reason ceases to perform its legitimate functions, the subjective mind usurps its place in this respect as in every other, and egotism in its most offensive sense and form is the inevitable result. These, then, are the distinctions which must be constantly borne in mind when considering this most important mental attribute of man.

That it *is* of the utmost importance, is manifest from the one fact alone, which has already been alluded to, that "egoism" bears the same relation to the psychical or spiritual entity that what is known to common observation as "the instinct of self-preservation" bears to the physical man. The one preserves the soul; the other preserves the body. It is, in fact, the one instinct of self-preservation manifested on both the physical and the spiritual planes. On the physical plane it has as many ways of manifesting itself as there are dangers threatening the safety of the body. On the spiritual plane there is but the one way for it to manifest itself; for there is but one possible danger that threatens the existence of the soul, and that is the loss of conscious personality.

The acts and motive power preservative of the physical

life are dynamic. The power preservative of the spiritual life is spiritual; it is the will, — the initial motive force or power belonging to mental or spiritual life.

Now, will being stimulated by desire, and the strongest desire of the soul being to retain its conscious personality, it follows that *the one great element of potency in the soul expends its greatest energy in an effort to retain its conscious personality independently of physical conditions.*

It will thus be seen that "egoism," which has heretofore been regarded as wholly abnormal, is, in reality, a faculty of the soul of infinite importance, having a normal function to perform which alone renders a continued existence in a future life possible. Egoism, as I have defined it, combined with faith, constitutes the mental dynamics of the soul. Without it the soul would be in much the same condition as the body would be in the absence of the instinct of self-preservation. Without the will or desire to live, without the instinct which causes the body to shrink from peril, one would be a passive victim to the myriad adverse forces of his environment. If he survived his childhood, it would be because of the watchful care of others. What would be the fate of the soul, *minus* its instinctive clinging to its personality, — *minus* the desire, the will, to retain its individuality and to sustain an independent existence as a distinct entity, — what, in short, would be its fate in the absence of the dynamics of the mind, can only be determined by logical deduction. Thus, if faculty implies function, the absence of faculty implies the absence of function. If the possession of legs implies the ability to walk, the absence of legs implies the antithesis.

The argument on this branch of the subject now stands thus: —

1. The essential prerequisites to the retention of personality are (*a*) Consciousness; (*b*) Memory; (*c*) Will.

2. The soul has a perfect memory, which performs no

normal function in this life, and is superfluous in a future life, considered merely as an aid to intellectual work, in view of its power of intuitional perception.

3. Corollary: Its ability to remember the facts of its experience can have no use or object other than that of the retention of its own personality and the recognition of other personalities.

4. Since memory presupposes consciousness, the latter must be presumed to be as perfect as the former.

5. Will constitutes the initial motive power of the human mind and soul.

6. Will has its biological origin in desire; and *egoism*, the strongest of human emotions, is the desire of the soul to retain its personality — to be saved from annihilation.

7. Corollary: The soul possesses all the mental powers necessary for the retention of its personality and for the maintenance of an existence independently of the body.

It will thus be seen that the subjective mind, or the mind of the soul, possesses all the mental attributes which constitute what may be termed the purely intellectual dynamics. It remains to consider another power, peculiarly its own, which demonstrates the actual possession by the soul of a kinetic potency which for the present cannot be classed as intellectual. I refer, of course, to its power to move ponderable bodies, otherwise known as the power of levitation. It is that power which, in spiritistic circles, produces raps upon floors, walls, and furniture, levitates the medium, tilts tables, and sometimes causes the most orderly and dignified parlor furniture to "cut fantastic tricks before high heaven." Of the existence of this force no one who has investigated the subject, pretends to doubt. It has been investigated by many of the ablest scientists of the world, notably by Prof. Elliot Coues, of Washington, and Professors Crookes and Lodge, of London, besides many other scientists of lesser note in Europe and America.

Professor Coues has given it a name, "Telekinesis," and writes on the subject learnedly and interestingly, as he writes on every subject which he handles; and Professor Crookes has given the world a very learned disquisition on the topic; whilst Professor Lodge has exhausted the resources of human ingenuity in devising tests demonstrative of the existence of the force, and of the English language in describing them. Spiritists, of course, have an easy way of accounting for it by referring it directly to spirits of the dead. But no scientist has been able to do more than to enable us to say that it is a power belonging exclusively to the subjective entity; that it performs no normal function in this life; that it requires a physical basis in order to produce phenomena cognizable by the objective senses, and thus, like all other psychic phenomena known as spiritistic, it is never produced except as a result of the most intensely abnormal physical and mental conditions. In reference to the claims of the spiritists it need only be said that there is no valid evidence whatever that disembodied spirits either do or can produce the phenomena of telekinesis.

The only thing that can be said of the power with certainty is that it exists; that it is not a power of the objective mind; that it is a power of the human soul, and that it is valuable in this life only as an evidence that there is a kinetic force resident in the soul. It will readily be seen, however, that the bare fact of its existence as a factor in the organism of the soul is of the most transcendent interest and importance, although we may never know in this life all that it implies when it is exercised in the future life. Nor is it important that we should know. But it is of the utmost importance that we should know that the soul is possessed of kinetic force, for it completes the chain of evidence necessary to demonstrate the fact that it possesses the power and potentiality of a self-existent entity. It is simply impossible to conceive of an intelligent organism *minus* kinetic force; for no matter how ethereal or imponderable the soul may be, measured by physical standards, it must be supposed to possess the power of moving, or causing motion, suitable to its environment. Besides, if, as we must suppose, the soul is a spark of the Divine Intelligence, it must be invested, in some degree, with the potential energy inhering in Omnipotence.

CHAPTER XX.

THE AFFECTIONAL EMOTIONS OF THE SOUL.

All the Affectional Emotions Retained in the Future Life. — Telepathy the Means of Communion in the Future Life. — Telepathy neither Vestigial nor Rudimentary. — It performs no Normal Function in this Life.

IN the chapters next preceding this I have shown, first, that in the faculty of intuitive perception of the laws of Nature the soul possesses the prerequisites of a purely intellectual existence; secondly, that, in its perfect memory combined with its egoism, it possesses both the desire and power to retain its personality or individuality; and, thirdly, that, in the abnormal phenomenon of telekinesis, the fact is demonstrated that the soul possesses that kinetic power which necessarily belongs to every intelligent organism. The possession of only these powers would characterize a purely intellectual being, retaining its personality, but divested of every affectional emotion and destitute of any means of intelligent communication with others. It is, perhaps, superfluous to remark that such a being would correspond exactly with the ideas of the Brahmans and Buddhists who believe that the destiny of man is absorption by the Deity, modified by retention of personality. It is notable that, in common with that other sect who hold that the absorption is complete and the personality lost, they take no account of the natural affections of mankind, and, consequently, ignore all possibility of a social life in Nirvana.

In point of fact, they hold that all the affectional emotions belong exclusively to the incarnate man, and hence they begin their preparations for Nirvana by crushing out every human affection or impulse that pertains to social or domestic life, retaining nothing of an emotional character save their own monstrous, monumental egotism.

Such an existence might gratify the aspirations of those who pass the greater part of their mortal life in that state of abnormal subjectivity of which egotism is the salient characteristic. But to the normal Occidental man, unused to the processes of self-hypnotization; unused to surrendering his reason, and plunging into the realm of subjective hallucination induced by auto-suggestion; who has been taught to regard the love of wife, children, and friends as among the purest and holiest emotions of the human soul, — to such a man the promise of a future life without the prospect of a reunion with the loved ones who have gone before or who are to follow after, would offer no attractions that he would not gladly exchange for annihilation. To the normal man or woman an existence without love or the capacity and means for social enjoyment would be worse than annihilation. It is only by a determined and persistent repression of the normal emotions, by means of an abnormal asceticism, that any human being can bring his mind to such a state of moral and affectional atrophy as to contemplate with equanimity a final separation from family, friends, and kindred.

Let us, then, still further examine the known attributes of the subjective mind, with a view of ascertaining whether there is any warrant for the assumption that in the future life we are to be bereft of all that we hold dear in this.

Fortunately we have not far to look; for, standing on the very threshold of the inquiry, is the broad and significant fact that all the emotions that impart joy or sorrow to humanity find their origin and seat in the subjective mind. It is true that their functions pertain in part to this life. It is true that their primary function is to perpetuate the species; that their normal activity gives life and light and love and joy and happiness to incarnate humanity; and that, perverted, they are the prolific source of sorrow, misery, degradation, and despair. Like every other attribute of the soul when uncontrolled by objective reason, that is, when perverted to base and ignoble uses, they are prolific of evil consequences; whilst their normal exercise is promotive of the highest good to humanity. But, whatever may be the result of their exercise here, the fact that the subjective mind is the seat of the emotions is demonstrative that they have a higher function to perform in a realm where perversion is impossible.

It will thus be seen that the love and affection which man bears to his fellow-man will not be blotted out of existence when the brain ceases to perform its functions; for it exists in that mind which performs its functions independently of the brain's existence, in that mind which grows stronger as the brain grows weaker, in that mind whose strongest observable manifestations occur in the hour of death. If there is no faculty without a function, it follows that the affectional emotions have a legitimate sphere of exercise in that home not made with hands. In other words, the existence of those emotions in the soul constitutes indubitable evidence that there will exist, in the life to come, ample means for their exercise; and that conclusion presupposes a reunion with the legitimate objects of our love.

There is now but one thing lacking in the attributes and powers of the soul to complete the mental equipment necessary for an enjoyable intellectual and social existence of the highest order conceivable by the mind of man. It is almost superfluous to say that the one other thing needful is a means of communication between disembodied souls, or to remark that this want finds an ample supply in the power of telepathy.

Telepathy, as has again and again been demonstrated, is a power belonging exclusively to the subjective mind; the objective mind does not possess it in the remotest degree. This fact is evidenced by every salient telepathic phenomenon. It is the subjective mind that reads, and it is the subjective mind that is read. The objective thoughts of one cannot be read by the subjective mind of another, unless the objective and subjective thoughts happen to be synchronous. Hence it is very rare that a telepathist reads what the sitter is consciously thinking of. These facts, however, are of such common knowledge that it would be a waste of time to enlarge upon them.

The important fact connected with telepathy is that it performs no normal function in this life. This is obvious from the fact that it is only under abnormal conditions of the body and mind that the phenomenon is observable.

Much ink has been wasted in discussing the question whether telepathy is vestigial or rudimentary. The fact is that there is not a scintilla of valid evidence to show that it is either. If it were vestigial, we should have the right to expect to find indubitable evidence of its existence in the lower animals. But the fact is that there is little evidence to show that they can communicate in that way with each other. They all have an oral or objective language of their own, and all their senses are infinitely more acute than man's. This would hardly be the case if telepathy existed in animals as a normal power capable of affording protection or contributing to their well being. I am not unmindful of the well-known experiment of Prof. C. V. Riley,[1] an eminent scientist of Washington, who occluded a foreign insect and released it two miles from its mate; and the two were found together the next morning. The learned professor has suggested telepathy as a possible explanation of the fact; but he would hardly regard it as conclusive evidence of the

[1] Since deceased.

existence of that power in insects, in view of the well-known sensory powers of many of the lower animals, including insects. Besides, it is a feat that is vastly outdone by the carrier pigeon, whose marvellous powers are referable entirely to the sense of sight. There is, however, much evidence to show that man can influence animals telepathically; but no conclusive evidence has yet been forthcoming to show that animals can so communicate with each other. Neither is there any evidence to show that man ever possessed the power of telepathy in any greater degree than he now possesses it, or that he was ever in a physical or mental condition more or less favorable to the development of that power than he is now. There is therefore no evidence whatever that the faculty is vestigial.

There is as little evidence that it is a "rudimentary sixth sense," as many learned men, who are fond of rudimentary speculations without facts, would have us believe. It is true that there are more telepathists now than ever existed before; and it is also true that there are more hysterical women, of both sexes, than ever existed before. Besides, telepathy has only recently been scientifically investigated, and the fact that it is a power of the human mind has only recently been demonstrated to the satisfaction of the scientific world. But the demonstration of a fact of such startling import has sent thousands into the field of experimental psychology, with the result that millions of experiments have been made, demonstrating nothing but the bare fact that the power exists, and that it cannot be made useful in this life. It has not advanced human knowledge one step in the direction of any useful result or in the development of any useful power. It would be difficult to show that, of all the experiments that have been made or of all the instances where it has been spontaneously manifested, there is one case where it has proved to be of any benefit whatever. In the very nature of things this must always be true, for the simple reason that the law of suggestion must always render every experiment uncertain until the result has been verified by objective means. No one who is aware of the existence of that law would ever dare to depend upon a telepathic message where any material interest was at stake; and until the law of suggestion can be nullified, that is to say, until all possible subjective hallucinations, arising from possible suggestions, can be eliminated as possible factors in supposed telepathic experiences, there can be no possible means of rendering telepathy useful in earthly life.

Again, if telepathy were either vestigial or rudimentary, it would be manifested under normal conditions. It would be equivalent to a contradiction in terms to suppose that a normal faculty must always be exercised under abnormal conditions. The only condition approaching normality under which telepathy is ever manifested is in dreams. But until the element of suggestion arising from waking thoughts or peripheral stimuli can be eliminated from dreams, it is obvious that they cannot be depended upon as sources of information in the affairs of this life.

No; telepathy performs no normal function on the physical plane. We can catch only occasional glimpses of it here, — just enough to enable us to know that in the future life, when physical organs of speech no longer exist, there is ample provision for intelligent communion with those who share our destiny.

CHAPTER XXI.

PRACTICAL CONCLUSIONS.

The Abnormality of Psychic Manifestations. — The Dangers attending Psychic Activity. — The Different Forms of Psychic Development. — Psychic Powers inversely Proportioned to Health. — Unsuspected Dangers. — Musicians. — Stenographers and Typewriters. — Compositors. — Genius and Insanity. — Opinions of Scientists. — Dr. MacDonald. — Summary. — Biographical Facts. — The Great Practical Lesson of Psychic Science. — Immorality, Vice, Crime, and Insanity the Result of Psychic Activity.

THE lessons which psychic science teaches pertain not alone to the future world, but they are of the utmost practical value in this life. Indeed, I speak the words of truth and soberness when I declare that there is no subject of human thought and investigation of such transcendent and imminent practical importance to mankind as that of psychic science. And the great lesson which it teaches, the lesson which embraces all the others, is that *psychic phenomena are never produced except under the most intensely abnormal conditions of the physical and the mental organism.*

It may sound paradoxical to say that man's most important study can be successfully prosecuted only under and by virtue of abnormal conditions; but it must be remembered that it has always been by means of the study of abnormal phenomena that much of what man knows of his normal conditions has been revealed to him.

If man were always physically and mentally normal, there would be comparatively little learned of his physical or mental structure, for there would be no incentive to study. He would then be like a perfect machine, which, so long as normal conditions prevail, may be successfully operated by one who knows little or nothing of its internal structure; but when the machine breaks, or for any reason fails to perform its normal functions, the operator is compelled to investigate the cause; and he thus becomes acquainted with its internal mechanism. It is because of disease that the physician becomes acquainted with the laws of health; and it is for the same reason that the surgeon is compelled to study the anatomy of the human frame. It was through the study of abnormal conditions of the body that Harvey was led to the discovery of the circulation of the blood. It was through the study of abnormal conditions of the mind that Hammond was led to the discovery that the brain is not the sole organ of the mind; and it was by means of abnormal conditions, congenital or induced, that he was enabled to demonstrate his theorem, and was thus enabled to give to the scientific world a physiological basis, not only for studying the problems of insanity, but upon which to postulate an immortal soul. In the physiological realm, abnormal conditions are practically the only incentives to serious study. In the psychical world they are at once the stimuli to the study of the science of the soul, and the means by which it can be successfully prosecuted.

I do not undertake to say that, because psychic phenomena are never produced except under abnormal conditions, they should never be produced. In view of the therapeutic value of hypnotism and cognate means of healing the sick, and especially in view of the precepts and example of Jesus, it would be absurd to attempt to prohibit the production of psychic phenomena. But the use of it for therapeutic purposes can be justified only on the same ground that the use of poisonous drugs for medicines can be justified. Their production can also be justified for purposes of scientific experiment; but only upon the same ground that we can justify experimental medicine or vivisection. It is absolutely necessary, then, that abnormal conditions be studied in order to enable us to understand and preserve normal conditions; and this is as true of the soul as it is of the body. But as it is unnecessary and improper for the physician to induce and perpetuate disease in a human body in order to study the laws of health, it is also in the highest degree improper, as well as unnecessary, to select a victim and continually induce an abnormal or diseased condition of his mind in order to study psychology. There is an abundance of disease both of body and of mind, existing all around us, from which all the necessary data can be obtained, without the necessity of immolating a human body or a human soul upon the altar of science. Nevertheless, it is probable that, for many years to come, experiments will be made with a view to new discoveries; and it is in the highest degree probable that hypnotism will, in some of its myriad forms, and when its laws are better understood, be largely employed not only for therapeutic purposes, but for moulding human character, especially in the young.

In the mean time the first great lesson to learn is that it is an abnormal condition at the best, and should never be tampered with by the ignorant; nor should it ever be employed except as a remedial agent, physical or moral, and then only by those who are familiar with its laws.

But the great lesson which psychic science teaches, and in which every human being is interested, is that all psychic activity is not only abnormal, but it is in the highest degree injurious to the body as well as to the mind.

I have again and again sought to impress upon my readers the pregnant fact that, whenever the subjective mind of man usurps control over the dual mental organism, Reason abdicates her throne; and, just in proportion to the completeness and persistency of that control, the subject is insane. Not only is this true, but it is a fact, which the experience of every-day life will demonstrate to the mind of the most superficial observer, that many of the nervous diseases to which mankind is subject, and *all immorality, vice, crime, and insanity are the direct results of abnormal psychic activity and control over the dual mental organization.*

Abnormal psychic activity in this life is the most insidious and formidable foe which man is called upon to encounter. It lurks everywhere and in every guise, and it threatens the stability of human society; for it strikes at the foundation of the most sacred of all human relations. It is the "drop of poison in man's purest cup." To substantiate this indictment, it is only necessary to call attention to a few of the leading classes of psychic phenomena, and note their effect, physical and moral, upon those who are the instruments of their production. It must be premised, however, that whilst it is true that psychic activity is destructive to the health of the body and utterly demoralizing to the nervous system, its untoward moral effects are largely due to utter ignorance, on the part of the psychic, of the source of the phenomena. Thus, spiritistic phenomena are always produced under the impression on the part of the psychic that they emanate from an extraneous source, — from a superior intelligence which the psychic can in no wise control; and the result is utterly demoralizing to the mind as well as to the body.[1]

No one who is familiar with the class of psychics known as "mediums" will gainsay the statement that, as a class, they are to the last degree neurotic. If there are exceptions to the rule, they are in cases where the psychic powers are of very recent development. It is probably true that

1 For a full discussion of this important point, see "The Law of Psychic Phenomena," ch. xxii.

perfectly healthy persons may develop psychic powers, although competent physicians are at variance on that point. The Charcot school of hypnotists hold that the psychic condition is itself a disease; whilst the Nancy school have seemingly demonstrated that perfectly healthy persons may be thrown into that condition. Be that as it may, the fact remains that the habitual indulgence in psychic practices, of any kind, grade, or character, will invariably result in some form of nervous derangement and disease; and if carried to excess, or continued long, will result in insanity or imbecility. Dr. Hammond, in his able work entitled "Spiritualism and Nervous Derangement,"[1] has forever settled the relations between psychic conditions and nervous diseases. They are always concomitant; and psychic activity may be either the cause or the result of nervous disease. But they invariably accompany each other. This is demonstrated by the facts, first, that no one can become a good psychic until a nervous derangement has been induced; second, the best psychics are those whose nervous systems are completely shattered; and, third, the degree of psychic power attainable by any one is in exact proportion to the intensity of his nervous derangement. The Seeress of Prevorst affords a striking example illustrating these propositions.[2] Bedridden for many years with a complication of the most terrible nervous disorders, she exhibited the most wonderful variety of psychic powers ever recorded. Mollie Fancher, the Brooklyn Enigma,[3] affords another striking illustration of the concomitance of nervous disease and psychic power. Owing to an accident which shattered her nervous system, this lady was thrown into a trance condition which lasted nine years, during which a series of the most wonderful psychic feats were

[1] Putnam's, N. Y., 1876.
[2] See "Seeress of Prevorst," London, 1845.
[3] See Judge Dailey's volume, Brooklyn, 1894.

performed. Her powers still continue to be manifested; but it is noted in the latter part of Judge Dailey's book that as she improves in health she is gradually losing her psychic power.

But it is useless to multiply citations of particular instances. It is a fact which can be verified by observation in any case where psychic phenomena are produced, that it requires an intensely abnormal condition of both body and mind to enable any one to achieve success in the psychic world; and this is true, without reference to the character of the manifestations, from the simplest telepathic experiment in an apparently normal condition down to the deepest trance that was ever induced in the passive victim of experimental hypnotism.

It is not, however, in the purely experimental part of psychic phenomena that the greatest and most insidious danger lurks. Like any other dangerous thing, it is all the more so when its presence is unsuspected. Psychic dangers are everywhere present, and exert their baleful influence in many places where the superficial observer would least expect to find them. Certain trades and occupations are beset with the danger in its most insidious form. For instance: —

Musicians are constantly in imminent danger of the evil consequences of an undue development of the subjective faculties; and their greatest peril arises from the fact that the foe is unsuspected. But a moment's consideration will render it plain that there are evils to guard against even in so noble an art as that of music. Music, as I have before pointed out, is the language of the emotions, and has its origin in the subjective mind. The mechanical execution of good music by a trained musician is automatic, and therefore subjective. The common experience of every musician worthy of the name will bear me out in this assertion; and no one will gainsay the statement that good musicians as a class are exceedingly emotional, and in the highest degree nervous, excitable, passionate, jealous. The simple reason is that they have cultivated the subjective faculties without knowing the laws pertaining to subjective mental activity, and, consequently, without realizing the dangers, physical, mental, and moral, which attend the undue development of the subjective faculties, uncontrolled by objective reason. I do not say that musicians, as a class, are more immoral than other classes of people, although that charge has been freely made by others who are, perhaps, actuated by unworthy motives; but I do say that they necessarily and habitually cultivate those faculties which, uncontrolled by reason, are liable to lead men astray. Fortunately, however, musicians are, as a rule, far above the average in point of mental cultivation and social standing, and hence are better qualified to appreciate the dangers which lurk in their pathway. They may, therefore, be able to guard against the moral evils attending the undue dominance of the subjective faculties and emotions, but can never wholly overcome the abnormal physical conditions incident to them.

Stenographers and Typewriters are also beset with dangers arising from the same source. I mean, of course, good stenographers and good typewriters; for good work in those lines is always the result of such training as is required to enable the operator to do the work automatically. When that amount of skill is acquired, the danger begins, for the obvious reason that automatism is purely subjective. No one acquainted with the facts will gainsay the observation that stenographers and typewriters, as a class, are the victims of nervous disorders, oftentimes approaching the verge of insanity. Numbers of them employed in the Government departments at Washington are compelled to abandon their positions, owing to the nervous strain to which they are subjected.

Compositors are also beset by the same dangers. Again, I mean good compositors, that is, those whose training in type-setting is such as to enable them to work automatically; and no really good compositor, that is, one who can do a big day's work with reasonable accuracy, ever does or ever can set type in any other way. Every one who is acquainted with the besetting weaknesses of the craft will indorse what I say, and will agree with me that their troubles and their habits are the result of abnormal subjective mental activity.

But it is useless to multiply instances. The whole principle may be summed up in the general statement that any employment which unduly develops the subjective powers in any direction whatever, is attended by abnormal physical and mental conditions. If any class of people deserve an ample compensation, and are entitled to shorter hours and increased pay, it is that class whose occupation results in the development of the subjective faculties. Moreover, if any deserve a rich reward in the world to come, it is those who resist the temptations which beset the pathway of those whose subjective faculties are unduly developed.

Another class of people whose subjective faculties are unduly developed deserves especial mention. I refer to men of genius. I have elsewhere[1] defined genius as being the result of the synchronous action of the two minds; a condition which confers upon the individual the benefit of the perfect memory and the other powers of the subjective mind, but does not deprive him of his objective powers of reason. A perfect synchronism of development and activity would necessarily produce a wonderful intelligence, an intellectual colossus, in whatever field of endeavor he might engage. History furnishes but one example of perfect synchronism of development, intellectual, moral, and religious, —Jesus of Nazareth. Shakespeare comes as near the

[1] See "The Law of Psychic Phenomena."

ideal of intellectual symmetry as any man of whom we have any means of judging. But we have numerous examples of genius where the synchronism was only partial. Great results have been achieved by them, and the world will never cease to cherish the memory and worship at the shrine of those intellectual giants whose works mark the epochs of history and of science and of the arts. Unfortunately, however, the personal history of men of genius reveals the fact that their intellectual development was abnormal to the last degree. So universally true is this that alienists find it difficult to distinguish the dividing line between genius and insanity, and are wont to couple the two in their treatises. Thus, Dr. Arthur MacDonald, Specialist in the Bureau of Education at Washington, in his able and intensely interesting work entitled "Abnormal Man,"[1] devotes a chapter to "Insanity and Genius," in which he summarizes a large number of facts relating to the personal and pathological characteristics of men of genius in all the ages of which history furnishes the data. It would be impossible to present the leading facts in fewer words, or in more concise and intelligent form, than Dr. MacDonald has summarized them. In view of the transcendent importance of the facts as bearing upon the subject under consideration, I shall take the liberty of making the following extracts from his report: —

"As an introduction to the biographical study of genius it will be interesting to give the opinions of geniuses themselves.

"Aristotle says that under the influence of a congestion of the head there are persons who become poets, prophets, and sibyls. Plato affirms that delirium is not an evil but a great benefaction when it emanates from the divinity.

"Democritus makes insanity an essential condition of poetry. Diderot says: 'Ah, how close the insane and the genius touch; they are imprisoned and enchained, or statues are raised to them.' Voltaire says: 'Heaven in forming us mixed our life with reason and insanity, the elements of our imperfect being;

[1] Circular of Information, No. 4, 1893, Bureau of Education.

they compose every man, they form his essence.' Pascal says: 'Extreme mind is close to extreme insanity.' Mirabeau affirms that common-sense is the absence of too vivid passion; it marches by beaten paths, but genius never. Only men with great passions can be great. Cato said, before committing suicide: 'Since when have I shown signs of insanity?' Tasso said: 'I am compelled to believe that my insanity is caused by drunkenness and by love; for I know well that I drink too much.' Cicero speaks of the *furor poeticus;* Horace of the *amabilis insania;* Lamartine of the mental disease called genius. Newton, in a letter to Locke, says that he passed some months without having a 'consistency of mind.' Chateaubriand says that his chief fault is weariness, disgust of everything, and perpetual doubt. Dryden says: 'Great wit to madness sure is near allied.' Lord Beaconsfield says: 'I have sometimes half believed, although the suspicion is mortifying, that there is only a step between his state who deeply indulges in imaginative meditation and insanity. I was not always sure of my identity or even existence, for I have found it necessary to shout aloud to be sure that I lived.' Schopenhauer confessed that when he composed his great work he carried himself strangely, and was taken for insane. He said that men of genius are often like the insane, given to continual agitation. Tolstoi acknowledges that philosophical scepticism had led him to a condition bordering on insanity. George Sand says of herself that at about seventeen she became deeply melancholic; that later she was tempted to suicide; that this temptation was so vivid, sudden, and bizarre that she considered it a species of insanity. Heine said that his disease may have given a morbid character to his later compositions.

"However paradoxical such sayings may seem, a serious investigation will show striking resemblances between the highest mental activity and diseased mind. As a proof of this we will give a number of facts, to which many more might be added.

"The difficulty of obtaining facts of an abnormal or pathological nature, and of other unfavorable data, is obvious. Authors have not only concealed such data, but have not deemed them important enough to record. It is due to the medical men, whose life brings them closest to abnormal reality, that such facts have been gathered. If it be said that the abnormal or exceptional must be taken with some caution, because it is natural for the mind to exaggerate striking characteristics, it must be remembered that such facts, when unfavorable to reputation, are concealed. In the study of any exceptional or abnormal individual, as the insane or the genius, one finds much more concealed than is known.

"Socrates had hallucinations from his familiar genius or dæmon. Pausanias, the Lacedemonian, after killing a young slave, was tormented until his death by a spirit, which pursued him in all places, and which resembled his victim. Lucretius was attacked with intermittent mania. Bayle says this mania left him lucid intervals, during which he composed six books, *De rerum natura*. He was forty-four years of age when he put an end to his life. Charles V. had epileptic attacks during his youth. He stammered. He retreated to a monastery, where he had the singular phantasy of celebrating his own funeral rites in his own presence. His mother (Jane of Castile) was insane and deformed. His grandfather (Ferdinand of Aragon) died at the age of sixty-two, in a state of profound melancholia. Peter the Great, during infancy, was subject to nervous attacks which degenerated into epilepsy. One of his sons had hallucinations; another, convulsions. Cæsar was epileptic, of feeble constitution, with pallid skin, and subject to headaches. Linné, a precocious genius, had a cranium hydrocephalic in form. He suffered from a stroke of paralysis. At the end of one attack he had forgotten his name. He died in a state of senile dementia. Raphael experienced temptations to suicide. He himself says: 'I tied the fisherman's cords which I found in the boat eight times around her body and mine, tightly as in a winding-sheet. I raised her in my arms, which I had kept free in order to precipitate her with me into the waves. . . . At the moment I was to leap, to be swallowed forever with her, I felt her pallid head turn upon my shoulder like a dead weight, and the body sink down upon my knees.'

"Pascal, from birth till death, suffered from nervous troubles. When he was only a year old, he fell into a languor, during which he could not see water without manifesting great outbursts of passion; and, still more peculiar, he could not bear to see his father and mother near each other. In 1627 he had paralysis from his waist down, so that he could not walk without crutches. This condition continued three months. During his last hours he was taken with terrible convulsions, in which he died. The autopsy showed peculiarities. His cranium appears to have no suture, unless, perhaps, the lambdoid or sagittal. A large quantity of the brain substance was very much condensed. Opposite the ventricles there were two impressions as of a finger in wax. These cavities were full of clotted and decayed blood, and there was, it is said, a gangrenous condition of the dura mater. Walter Scott, during his infancy, had precarious health, and before the age of two, was paralyzed in his right limb. He had a stroke of apoplexy. He had this vision on hearing of the death of Byron: Coming into the dining-room, he saw before him the image of his dead friend. On advancing toward it, he recognized that the vision was due to drapery extended over the screen.

"Some men of genius who have observed themselves describe their inspiration as a gentle fever, during which their thoughts become rapid and involuntary. Dante says: —

' . . . I' mi son un che, quando
Amore spira, noto ed in quel modo
Che detta dentro vo significando.'

(I am so made that when love inspires me, I attend; and according as it speaks in me, I speak.)

"Voltaire, like Cicero, Demosthenes, Newton, and Walter Scott, was born under the saddest and most alarming conditions of health. His feebleness was such that he could not be taken to church to be christened. During his first years he manifested an extraordinary mind. In his old age he was like a bent shadow. He had an attack of apoplexy at the age of 83. His autopsy showed a slight thickness of the bony walls of the cranium. In spite of his advanced age, there was an enormous development of the encephalon.

"Michael Angelo, while painting 'The Last Judgment,' fell from his scaffold and received a painful injury in the leg. He shut himself up and would not see any one. Bacio Rontini, a celebrated physician, came by accident to see him. He found all the doors closed. No one responding, he went into the cellar and came upstairs. He found Michael Angelo in his room, resolved to die. His friend the physician would not leave him. He brought him out of the peculiar frame of mind into which he had fallen. The elder brother of Richelieu the Cardinal was a singular man; he committed suicide because of a rebuke from his parents. The sister of Richelieu was insane. Richelieu himself had attacks of insanity; he would figure himself as a horse, but afterwards would have no recollection of it. Descartes, after a long retirement, was followed by an invisible person, who urged him to pursue his investigations after the truth. Goethe was sure of having perceived the image of himself coming to meet him. Goethe's mother died of an apoplectic attack. Cromwell, when at school, had a hallucination in his room; suddenly the curtains opened and a woman of gigantic stature appeared to him, announcing his future greatness. In the days of his power he liked to recount this vision. Cromwell had violent attacks of melancholic humor; he spoke of his hypochondria. His entire moral life was moulded by a sickly and neuropathical constitution, which he had at birth. Rousseau was a type of the melancholic temperament, assuming sometimes the symptoms of a veritable pathetic insanity. He sought to realize his phantoms in the least susceptible circumstances; he saw everywhere enemies and conspirators (frequent in the first stage of insanity). Once coming to his sailing-vessel in England, he interpreted the unfavorable winds as a conspiracy against him; then mounted an elevation and began to harangue the people, although they did not understand a word he said. In addition to his fixed ideas and deliriant convictions, Rousseau suffered from attacks of acute delirium, a sort of maniacal excitation. He died from an apoplectic attack. Jeanne d'Arc was a genius by her intrepid will; she had faith in her visions; her faith rested upon the immovable foundation of numerous hallucinations having the force of moral and intellectual impulsion, making her superior to those around her. Science can pronounce as to her inspirations, but its judgment does not diminish in the least the merit of her heroism. Jeanne was of the peasant class and uneducated. According to her statement, she first heard supernatural voices when she was 13 years old. Mohammed was epileptic. He persistently claimed to be a messenger from God, receiving his first revelation at the age of 42. He lost his father in infancy and his mother in childhood; was a travelling merchant, and married a wealthy widow fifteen years older than himself. His revelations began with visions in sleep. He used to live alone in a cave. He had interviews with the angel Gabriel. Henry Heine died of a chronic disease of the spinal column. Lotze was often melancholic. Molière suffered from convulsions; delay or derangement could throw him into a convulsion.

"Mozart's musical talent was revealed at three years of age; between four and six he composed pieces with expertness. Mozart died at 36 of cerebral hydropsy. He had a presentiment of his approaching end. He was subject to fainting fits before and during the composition of his famous 'Requiem.' Mozart always thought that the unknown person which presented itself to him was not an ordinary being, but surely had relations with another world, and that he was sent to him to announce his end. Cuvier died of an affection of the nervous centres; the autopsy showed a voluminous brain. He lost all his children by a fever called 'cerebral.' Condillac had frequent attacks of somnambulism; he sometimes found his work finished in the morning. Bossuet suffered from a disease from which he once lost speech, knowledge, and even the faculty of understanding. Dumas says: 'Victor Hugo was dominated by the fixed idea to become a great poet and the greatest man of all countries and times. For a certain time the glory of Napoleon haunted him.' Chopin ordered by will that he be buried in a gala costume, white cravat, small shoes, and short trousers. He abandoned his wife, whom he loved, because she offered another person a seat before she offered it to him. Giordano Bruno considered himself enlightened by a superior light sent from God, who knows the essence of things. Comte considered himself the 'Great Priest' of humanity. Madame de Staël died in a state of delirium, which had lasted several days; according to some authors, several months. The autopsy showed a large quantity of cerebral matter, and very thin cranium. Moreau of Tours said she had a nervous habit of rolling continually between her fingers small strips of paper, an ample provision of which was kept on her mantelpiece. She used opium immoderately. She had a singular idea during her whole life; she was afraid of being cold in the tomb; she desired that she be enveloped in fur before burial.

"English men of letters who have become insane, or have had hallucinations and peculiarities symptomatic of insanity, are Swift, Johnson, Cowper, Southey, Shelley, Byron, Goldsmith, Lamb, and Poe. Swift was also cruel in conduct, but he was hardly responsible, as his insanity was congenital. His paternal uncle lost speech and memory, and died insane. Swift was somewhat erratic and wild as a university student. He suffered at times from giddiness, impaired eyesight, deafness, muscular twitchings, and paralysis of the muscles on the right side of the mouth. He had a bad temper, was called 'mad person;' actually feared insanity, saying once, on seeing a tree that had been struck by lightning, 'I shall be like that tree: I shall die at the top.' Later in life he became a violent maniac. The post-mortem examination showed a cerebral serous effusion and softening of the cortex. There were a number of cranial anomalies. Shelley, when young, was strange and fond of musing alone, and was called 'Mad Shelley;' he suffered from somnambulism and bad dreams, and was excitable and impetuous; these symptoms increased with age; at twenty he constantly took laudanum for his nervous condition; he had hallucinations; he saw a child rise from the sea and clasp his hands, a vision which it was difficult to reason away. Much eccentricity existed in the immediate antecedents of Shelley. Charles Lamb was confined in an insane asylum. Johnson was hypochondriacal and apprehended insanity, fancying himself seized with it; he had convulsions, cramps, and a paralytic seizure depriving him of speech; he had hallucinations of hearing. Carlyle considered Southey the most excitable man of his acquaintance. Southey's mind failed, and he became an imbecile and died; a year before his death he was in a dreamy state, little conscious of his surroundings. Southey wrote verses before he was eight years of age. His maternal uncle was an idiot, and died of apoplexy. The mother of Southey had paralysis. Cowper was attacked with melancholia at twenty, which continued a year; at another time it returned with greater force. He himself tells of his attempts at suicide; he bought laudanum, keeping it in his pocket, when later a feeling pressed him to carry it into execution; but soon another idea came to him, to go to France and enter a monastery; then the suicidal impulse came again, to throw himself into the river, — an inhibitory feeling from taking the laudanum, — but he would have succeeded in hanging himself had not the thong to which the rope was fastened broken. After suicidal ideas left him, he relapsed into religious melancholia, thinking he had committed the unpardonable sin. He was confined in an asylum eighteen months. Keats was an extremely emotional child, passing from laughter to tears: he was extremely passionate, using laudanum to calm himself; sometimes he fell into despondency. He prophesied truly that he would never have any rest until he reached the grave. The attacks of critics agitated him almost to insanity. His nervousness was very susceptible, so that even 'the glitter of the sun' or 'the sight of a flower' made his nature tremble. Coleridge was a precocious child, self-absorbed, weakly, and morbid in imagination; this morbidity was the cause of his running away from home when a child and from college when a student; he enlisted as a soldier, and again went to Malta for no reason, permitting his family to depend upon charity. When thirty years of age, his physical suffering led him to use opium. Subsequently he had a lateral curvature of the spine (De Quincey). There were many morbid symptoms in the family. Burns says: 'My constitution and frame were *ab origine* blasted with a deep, incurable taint of melancholia, which poisons my existence.' Dickens died from effusion of blood upon the brain; he was a sickly child, suffering from violent spasms; when a young man, he had a slight nervousness which increased with age, and finally was attacked with incipient paralysis. George Eliot suffered from melancholic moods, and from her thirtieth year had severe attacks of headache. As a child she was poor in health, and extremely sensitive to terror in the night. She remained a 'quivering fear' throughout her whole life. De Quincey, the opium-eater, took opium as a relief from neuralgia and general nervous irritation. He was in bad health for a long time, dying at the age of seventy-four. Alfred de Musset had attacks of syncope; he died at forty-seven. George Sand described him in the Forest of Fontainebleau in his neurotic terror, in his joy and despair, as manifesting a nervous condition approaching delirium. He had a morbid cerebral sensibility, showing itself in hallucinations; he had a suicidal inclination. He was a dissipated gambler, passing from gayety to depression. His keen disappointment in love in Italy was accompanied by brain fever. For some time after this he could not speak of his chagrin without falling into syncope. He had an hallucination, and to distinguish it from real things he had to ask his brother. Wellington was subject to fainting fits; he had epilepsy and died from an attack of the disease. Warren Hastings was sickly during his whole life; in his latter years he suffered from paralysis, giddiness, and hallucinations of hearing. During the time of his paralysis he developed a taste for writing poetry. Carlyle, the dyspeptic martyr, showed extreme irritability. He says in his diary: 'Nerves all inflamed and torn up, body and mind in a hag-ridden condition.' He suffered from a paralysis in his right hand. Carlyle's antecedents were conspicuously of a nervous kind. Bach died from a stroke of apoplexy; one of his numerous children was an idiot. His family suffered from nervous diseases. Handel was very irritable; at the age of fifty he was stricken with paralysis, which so affected his mind that he lived in retirement for a year.

"Nisbet says: 'Pathologically speaking, music is as fatal a gift to its possessor as the faculty for poetry or letters; the biographies of all the greatest musicians being a miserable chronicle of the ravages of nerve disorder extending, like the

Mosaic curse, to the third and fourth generation.' Newton in the last years of his life fell into a melancholia which deprived him of his power of thought. Newton himself, in a letter to Locke, says that he passed some months without having 'a consistency of mind.' He was also subject to vertigo. From the manner of manifestation and the results following from this disease, Moreau goes so far as to say that it permits a certain degree of diagnosis and may be called acute *dementia.*

"The insanity of Tasso is probable from the fact that, like Socrates, he believed he had a familiar genius which was pleased to talk with him, and from whom he learned things never before heard of. Swift died insane. Chateaubriand during his youth had ideas of suicide, and attempted to kill himself. His father died of apoplexy; his brother had an eccentricity bordering on insanity; was given to all vices and died of paralysis. 'My chief fault,' says Chateaubriand, 'is weariness, disgust of everything, and perpetual doubt.' Tacitus had a son who was an idiot. Beethoven was naturally bizarre and exceedingly irritable. He became deaf, and fell into a profound melancholia, in which he died. Alexander the Great had a neurosis of the muscles of the neck, attacking him from birth, and causing his head to incline constantly upon his shoulders, He died at the age of thirty-two, having all the symptoms of acute delirium tremens. His brother Arrchide was an idiot. His mother was a dissolute woman; his father was both dissolute and violent. De Balzac (Honoré) died of hypertrophy of the heart, a disease that can predispose one to cerebral congestion. The eccentricity of his ideas is well known. Lamartine says he had peculiar notions about everything; was in contradiction with the common-sense of 'this low world.' His father was as peculiar. Lord Chatham was from a family of original mental disproportions, of peculiarities almost approaching alienation. Lord Chatham did not do things as others: he was mysterious and violent, indolent and active, imperious and charming. Pope was rickety. He had this hallucination: one day he seemed to see an arm come out from the wall, and he inquired of his physician what this arm could be. Lord Byron was scrofulous and rachitic and club-footed. Sometimes he imagined that he was visited by a ghost; this he attributed to the over-excitability of his brain. He was born in convulsions. Lord Dudley had the conviction that Byron was insane. The Duke of Wellington died of an apoplectic attack. Napoleon I. had a bent back; an involuntary movement of the right shoulder, and at the same time another movement of the mouth from left to right. When in anger, according to his own expression, he looked like a hurricane, and felt a vibration in the calf of his left leg. Having a very delicate head, he did not like new hats. He feared apoplexy. To a general in his room he said, 'See up there.' The general did not respond. 'What,' said Napoleon, 'do you not discover it? It is before you, brilliant, becoming animated by degrees; it cried out that it would never abandon me; I see it on all great occasions; it says to me to advance, and it is for me a constant sign of fortune.'

"Originality is very common, both to men of genius and the insane; but in the latter case it is generally without purpose. Lombroso goes so far as to make unconsciousness and spontaneity in genius resemble epileptic attacks. Hagen makes irresistible impulse one of the characteristics of genius, as Schüle does in insanity. Mozart avowed that his musical inventions came involuntary, like dreams, showing an unconsciousness and spontaneity which are also frequent in insanity. Socrates says that poets create, not by reflection, but by natural instinct. Voltaire said, in a letter to Diderot, that all manifestations of genius are effects of instinct, and that all the philosophers of the world together could not have given 'Les animaux malades de la peste,' which La Fontaine composed without knowing even what he did."

The remark of Voltaire, above quoted, was itself an inspiration; for it furnishes the key to the whole subject. "All manifestations of genius," says he, "are the effects of instinct," — that is to say, all manifestations of genius are the results of cultivation of the subjective faculties; and all the abnormalities of genius are the results of the predominance of the subjective faculties over the faculties of objective reason and judgment.

It is obvious that if there is any one form of psychic development that is useful to mankind, it is that of genius; and it is equally obvious that if there is any one form of psychic development that could possibly be harmless to the physical organism, it must be in cases where the objective and subjective faculties are developed in more or less perfect synchronism. As genius affords the best, nay, the only illustrations of the most useful and at the same time the least harmful of all manifestations of psychic activity, I have ventured to avail myself of the researches of one of the most eminent students of the abnormal in mankind for the purpose of showing that there is but one step between insanity and the least harmful of psychic manifestations.

One of the great practical lessons, therefore, which psychic science teaches is that, normally, this is an objective world, — the realm of physical life and activity. God has endowed us with faculties of mind exactly fitted for our physical environment; and they are all-sufficient to enable us to master the forces of physical Nature so far as to render our brief sojourn within its realm tolerable and even pleasant. Those are the faculties, therefore, which we should cultivate in this form of existence; for their functions pertain exclusively to this life, and to no other. On the other hand, psychic science teaches us that we are the possessors of other faculties which perform no normal functions in this life; and practical experience shows that the habitual exercise of those faculties in this life produces the most disastrous results to both body and mind.

The conclusion is irresistible that we should carefully refrain from exercising and developing, in this life, those powers which belong exclusively to another form of existence; and the necessity for this inhibition becomes still more apparent when we remember that all immorality, all vice, all crime, and all insanity arise from one and the same cause, namely, the dominance of the subjective faculties; and that all exercise of psychic powers for other than works of necessity, and all practices which develop and cultivate the subjective faculties, have a direct tendency to arouse to abnormal activity those emotions and propensities which, uncontrolled by reason, lead to immorality, vice, crime, and insanity.

CHAPTER XXII.

LOGICAL AND SCIENTIFIC CONCLUSIONS.

A Perspective View of the Arguments Presented. — The Final Syllogism. — The Parable of the Rich Man and Lazarus. — The Christian's Heaven. — The Revelations of Modern Science Identical with those of Jesus.

I HAVE now briefly outlined a few of the principal arguments for a future life which are based upon the observable and demonstrable facts of experimental psychology, so far as those facts have been definitely ascertained through modern scientific investigations. The treatment has necessarily been brief; for, although the science of the soul is yet in its infancy, the basic facts have accumulated at an astonishing rate since the world has learned where to look for them. From the great mass of data thus far available I have selected what seemed to be the most important, and, to borrow a phrase from art, I have delineated them thus far in sectional detail. A perspective view will now be attempted in the form of a brief *résumé* of the salient features of the argument. This will be done at the risk of what might be considered unnecessary repetition; but the intelligent reader will agree with me that fundamental facts and principles cannot be too thoroughly impressed upon the mind of the earnest and conscientious searcher after truth.

The fundamental axiom upon which our argument is based, and which the reader is again requested constantly to bear in mind, is this: There is no faculty, emotion, or organism of the human mind that has not its use, function, or object.

The first great fundamental fact presented to view is that man is endowed with a dual mind. This has been abundantly demonstrated by the facts of experimental hypnotism, cerebral anatomy, and experimental surgery. It has also been shown to be a primordial fact of psychic evolution.

The fact of duality alone, considered in connection with our fundamental axiom, is sufficient to put the intelligent observer upon an earnest inquiry into the possible use, function, and object of a dual mental organism; and his first inquiry is, "What possible use is there for two minds if both are to perish with the body?" A future life, therefore, is at once suggested by this one isolated fact; and the suggestion is further strengthened by the fact that, whilst one of the two minds grows feeble as the body loses its vitality and is extinguished when the brain ceases to perform its functions, the other mind grows strong as the body grows weak, stronger still when the brain ceases to act, and reaches its maximum of power to produce observable phenomena at the very hour of physical dissolution. It is simply impossible, from these two facts alone, to resist the conclusion that the mind which reaches its maximum of observable power at the moment of dissolution is not extinguished by the act of dissolution. These facts, therefore, constitute presumptive evidence of a future life. They are not claimed to be conclusive; yet it can truly be said that men of sound judgment habitually stake their dearest interests upon evidence less demonstrative of vital propositions. It would, indeed, be difficult to find any other rational hypothesis that would explain all the phenomena pertaining to these two facts.

The next great fact, or congeries of facts, which presents itself to view is that —

1. Each of the two minds possesses powers and functions which are not shared by the other.

2. Each of the two minds is hedged about by limitations not shared by the other.

3. These powers and limitations are divided into three distinct classes; namely, —

(*a*) Those which belong exclusively to the objective mind;

(*b*) Those which belong exclusively to the subjective mind;

(*c*) Those which are common to both minds.

4. Those which belong to class (*a*) pertain exclusively to physical life and environment.

5. Those which belong to class (*b*) perform no function whatever in physical life, and are observable only under abnormal physical conditions.

6. Those which belong to class (*c*) are more or less imperfect — finite — in their manifestations in the objective mind, whereas *each faculty is perfect in the subjective mind.*

Thus we find man, as he is presented to us in the light of demonstrable facts, possessed of a dual mental organism, comprising two classes of faculties, each complete in itself.

We find one class of faculties to be finite, perishable, imperfect, and yet well adapted to a physical existence and a material environment, and capable of development, by the processes of evolution, to a high degree of excellence, morally, physically, and mentally, within the limits of its finite nature. We also find that the noblest faculties belonging to physical man — those faculties which alone render his existence in this life tolerable or even possible, those faculties which give him dominion over the forces of physical nature — are faculties which pertain exclusively to this life.

On the other hand, we find another set of faculties, each perfect in itself, and complete in the aggregate, — that is to say, every faculty, attribute, and power necessary to constitute a complete personality being present in perfection; and we find that the most important of those faculties perform no normal function in physical life.

Here, then, we have a personality, connascent with the physical organism, but possessing independent powers; a distinct entity, with the intellect of a god; a human soul, filled with human emotions, affections, hopes, aspirations, and desires; longing for immortal life with a passionate yearning that passeth understanding; possessing, in a word, all the intellectual and moral attributes of a perfect manhood, together with a kinetic force often transcending, in its visible manifestations, the power of the physical frame; in a word, "a perfect being, nobly planned," — a being of godlike powers and of infinite possibilities.

In his apostrophe to man, Shakespeare must have embodied a description of an inspired vision, not of a mere human entity as it is visible in the flesh, but of a disembodied soul, clothed with the investiture of Heaven, and in full possession of its heritage of immortal attributes. It was a dream of such a being that he put into the mouth of Hamlet in these memorable words: —

> "What a piece of work is a man! how noble in reason! how infinite in faculty! in form and moving how express and admirable! in action how like an angel! in apprehension how like a god!"

The reader will not fail to remember the last exclamation in connection with what has been said of man's powers of intuitional perception of divine truth, — a power which belongs alone to the soul; a power which in itself is demonstrative of kinship to God, because it is the essential attribute of Omniscience.

Is it conceivable that there has been created such a manhood without a mission, such faculties without a function, such powers without a purpose?

Impossible! If Nature is constant, no faculty of the human mind exists without a normal function to perform. If no faculty exists without a normal function to perform, those faculties which do exist must perform their functions either in this life or a future life. If man possesses faculties which perform no normal function in this life, it follows that the functions of such faculties must be performed in a future life.

Or, to put the argument in a still more concise and purely syllogistic form, the propositions stand thus: —

Every faculty of the human mind has a normal function to perform either in this life or in a future life.

Some faculties of the human mind perform no normal functions in this life.

Therefore, *Some faculties of the human mind are destined to perform their functions in a future life.*

No scientist will for a moment question the soundness of the major premise of the above syllogism. It is self-evident, — axiomatic.

No one who is at all familiar with the results of modern scientific research in the field of psychic phenomena will for a moment gainsay the minor premise. The one faculty of telepathy alone is demonstrative of the soundness of that proposition, to say nothing of the faculty of intuitive perception, etc.

The major and minor premises being each demonstrably true, the soundness of the conclusion that man is destined to inherit a future life is self-evident.

It will be observed that, in constructing this final syllogism, I have done so without reference to the theory of the dual mind. Not that I have the slightest doubt of the scientific accuracy of that hypothesis; for I can have none in view of the array of facts which have been presented. But, as I have already pointed out, the theory of a dual mind and the theory of a unitary mind with dual faculties are concurrent hypotheses, and lead to identical conclusions. Hypothesis is not a final dogma: it is merely an instrument of logic. It is the divining-rod of truth. Facts are primordial, antecedent, and ultimate, and exist independently of any hypothesis that may be employed to account for them. The fact that the human mind is endowed with two distinct classes of faculties is demonstrable in itself, and exists independently of either the dual or the unitary hypothesis. That being the essential fact, I have framed my syllogism in terms broad enough to arrest the attention and extort the assent of the scientist who is not yet ready fully to indorse the dual hypothesis.

I have now finished my argument for a future life. If the facts which have been adduced do not demonstrate my thesis, crudely and imperfectly as they have been presented, then Nature herself has performed a miracle, and demonstrated her inconstancy.

Before closing, however, I desire to draw attention to one general conclusion, derivable from the facts herein presented, which must be a source of pride and gratulation to every inhabitant of Christian lands, whatever may be his individual belief or bias on the subject of Christianity. That conclusion is that the facts of psychic science fully and completely sustain the religious philosophy of Jesus of Nazareth, demonstrate his perfect mastery of the science of the soul, and confirm every essential doctrine of the Christian religion. It is almost superfluous to remark that this can be said of no other religion on earth. It is true that the religious philosophy of the Hindu is founded upon an observation of the same psychic phenomena. But, as I have remarked in previous chapters, their point of observation did not take in the whole field,—that is, they did not take into consideration all of the powers and attributes of the soul; nor did their partial observation bear the stamp of scientific accuracy, owing to their ignorance of the fundamental law of psychic science. Whereas the Christian religion is based upon an accurate survey of the entire field of psychic science by the most colossal religious genius the world has ever seen. Not only was the whole field surveyed by him, but it was with a full and accurate intuitive knowledge of every principle involved, as well as of every attribute of the human soul. That this is true, is scientifically demonstrated by the fact that modern scientific induction reveals, in every detail, the same truths which Jesus proclaimed eighteen hundred years ago.

The most specific utterance of Jesus concerning the future life and its conditions is contained in the parable of the rich man and Lazarus, which reads as follows: —

"There was a certain rich man, which was clothed in purple and fine linen, and fared sumptuously every day:

"And there was a certain beggar named Lazarus, which was laid at his gate, full of sores,

"And desiring to be fed with the crumbs which fell from the rich man's table: moreover the dogs came and licked his sores.

"And it came to pass, that the beggar died, and was carried by the angels into Abraham's bosom: the rich man also died, and was buried;

"And in hell he lifted up his eyes, being in torments, and seeth Abraham afar off, and Lazarus in his bosom.

"And he cried and said, Father Abraham, have mercy on me, and send Lazarus, that he may dip the tip of his finger in water, and cool my tongue; for I am tormented in this flame.

"But Abraham said, Son, remember that thou in thy lifetime receivedst thy good things, and likewise Lazarus evil things: but now he is comforted, and thou art tormented.

"And beside all this, between us and you there is a great gulf fixed: so that they which would pass from hence to you cannot; neither can they pass to us, that would come from thence.

"Then he said, I pray thee therefore, father, that thou wouldest send him to my father's house:

"For I have five brethren; that he may testify unto them, lest they also come into this place of torment.

"Abraham saith unto him, They have Moses and the prophets; let them hear them.

"And he said, Nay, father Abraham: but if one went unto them from the dead, they will repent.

"And he said unto him, If they hear not Moses and the prophets, neither will they be persuaded, though one rose from the dead." [1]

This parable, couched as it is in general terms, conveys, nevertheless, specific information of the greatest importance. The words "And in hell he lifted up his eyes" and saw "Abraham afar off" convey the information that the souls of men will recognize each other in the future life.

The expression "And he cried and said, Father Abraham," etc., shows that spirits communicate with each other in the other world.

The words "But Abraham said, Son, remember that thou in thy lifetime," etc., tell us that we remember the particulars of our earthly life; whilst the remaining part of the same clause teaches us, generally, that it is through our memory that we are punished for the deeds done in the body; and, specifically, that if the spirit of charity and brotherly love is not regnant in our breasts in this life, the memory of our neglect to relieve human suffering will be a source of torment to us in the world to come. It is nowhere stated that the rich man was not a good citizen in the ordinary affairs of life. The only charge against him was that out of his abundance of this world's goods he failed to relieve the distress of the beggar at his door.

The words "Then he said, I pray thee therefore, father, that thou wouldest send him to my father's house: for I have five brethren," etc., conveys, in unmistakable language, the information that, good or bad as we may be in this life, we retain, in the future life, our affection for those we love in this form of existence; and that it is partly through our affectional emotions that we are made happy or wretched in the life to come.

The unmistakable import of the closing clauses of the parable is that it is neither expedient nor possible, for any

[1] Luke xvi. 19–31.

purpose whatever, for spirits of the dead to communicate with the living.

This parable was obviously intended to convey to the world the sum total of all the knowledge of the conditions of the future life which Jesus could successfully impart to the finite comprehension of his followers. If communication of spirits of the dead with the living is a possibility, he would have taken that occasion to impart the information.

Taking it for granted that Jesus knew the laws of the soul, and was aware of its powers, functions, and limitations, it is impossible, without impugning his character for sincerity, to suppose that he could have uttered the words of the closing sentences of that parable if spirit communication with the living is either expedient or possible. The points of information, however, which he did from time to time impart, are of the utmost value and importance, and, moreover, they coincide exactly with the inductions of modern science.

Thus, modern psychic science reveals the same omnipotent, omnipresent, omniscient, immanent God, — the same loving, tender, benevolent, merciful Father whom Jesus was the first to proclaim.

It reveals the same frail man, possessing all the powers, attributes, and limitations which Jesus declared or exemplified.

It reveals the same immortal destiny for man which Jesus "brought to light," and prescribes the same conditions precedent to its enjoyment.

It reveals the soul of man as the possessor of all the faculties, affections, and emotions which are requisite and necessary for the enjoyment of perfect felicity in the future life.

And it also reveals, in those same faculties and affectional emotions, — in the perfect memory of every detail of the acts and deeds of earthly life, together with the awakened conscience resulting from the intuitive perception of the eternal principles of right and wrong, — a most perfect means for conferring the rewards promised by the Christian religion for a well-spent life, as well as for meting out the punishments for vice and crime, in exact and necessary accordance "with the deeds done in the body."

Moreover, science ceases its revelations at the very point where Jesus paused; namely, at the portals of the tomb. He gave us an assurance of a future life; and science confirms his words. He assured us of abundant rewards in the future life for righteousness in this; and science reveals in us the capacity for the enjoyment of the promised rewards. Beyond that his lips were sealed. Beyond that science cannot penetrate.

THE END.

www.ingramcontent.com/pod-product-compliance
Lightning Source LLC
Chambersburg PA
CBHW021501210326
41599CB00012B/1084
9783337419660